Universitext

For further volumes:
http://www.springer.com/series/223

René Schoof

Catalan's Conjecture

Springer

René Schoof
Università di Roma
Tor Vergata
Italia
schoof@mat.uniroma2.it

ISBN: 978-1-84800-184-8   e-ISBN: 978-1-84800-185-5
DOI: 10.1007/978-1-84800-185-5

British Library Cataloguing in Publication Data
A catalogue record for this book is available from the British Library

Library of Congress Control Number: 2008933674

Mathematics Subject Classification (2000): 11D41; 11D61; 11R18

Springer Science+Business Media
springer.com

Preface

A first draft of this book was written in August 2003. It was based on two sets of lecture notes by Yuri Bilu. In the following years, the draft was used in student seminars in Rome, on the island of Minorca, and in Regensburg. With the benefit of these experiences, a new version was written. It is the basis of this book.

The notes contain a complete proof of Catalan's conjecture. To read the first few chapters requires little more than a basic mathematical background and some knowledge of elementary number theory. The other chapters involve Galois theory, some more algebraic number theory, and a little bit of commutative algebra [1]. The basic facts from the arithmetic of cyclotomic fields are discussed in the text. This material can also be found in the textbooks *Introduction to Cyclotomic Fields* by L.C. Washington [50] and *Cyclotomic Fields* by S. Lang [25].

Our exposition is self-contained with one small exception. This regards chapter 14. Here we explain an argument of Mihăilescu's that is based on Francisco Thaine's famous theorem. Our proof of Thaine's theorem involves an application of Chebotarëv's density theorem to the Hilbert class field of a cyclotomic field. While we do provide a proof of Chebotarëv's theorem, we do *not* prove the existence and the basic properties of the Hilbert class field. A proof would involve a good deal of *class field theory*, and this is not included in these notes. We hope instead that the present application motivates the interested reader to study class field theory. This theory is exposed in the textbooks Algebraic Number Theory by J.W.S. Cassels and A. Fröhlich [10] and Algebraic Number Theory by S. Lang [24].

I would like to thank Leonardo Cangelmi, Alessandro Conflitti, Jeanine Daems, Carlo Gasbarri, Danny Gomez, David Kohel, Hendrik Lenstra, Preda Mihăilescu, Peter Stevenhagen, Andrea Susa, Michiel Vermeulen, Valerio Talamanca, Michael Tse, Filippo Viviani, Larry Washington, and Gabor Wiese for their remarks on earlier versions of these notes. I especially thank Yuri Bilu, whose 2002 manuscript [3] and 2003 Oberwolfach lecture [4] were very useful to me.

Rome   René Schoof

May 2007

Leitfaden

If you know algebraic number theory and the theory of cyclotomic fields and are only interested in Mihăilescu's proof of Catalan's conjecture, read chapters 1, 7, 8, 10, 11, 12, and 14. If you want to see a proof *from scratch* of Catalan's conjecture, then also read chapters 2, 3, 4, and 6. Chapters 5, 9, 13, 15, and 16 deal with some of the more advanced prerequisites. Here we discuss Runge's method, Stickelberger's theorem, basic properties of semisimple group rings, the Chebotarëv density theorem, and Thaine's theorem, respectively.

Contents

Contents

1
Introduction

In this book, we present Preda Mihăilescu's beautiful proof of the conjecture made by Eugène Charles Catalan in 1844 in a letter [11] to the editor of Crelle's journal:

> Je vous prie, Monsieur, de vouloir bien énoncer, dans votre recueil, le théorème suivant, que je crois vrai, bien que je n'aie pas encore réussi à le démontrer complètement: d'autres seront peut-être plus heureux:
>
> Deux nombres entiers consécutifs, autres que 8 et 9 ne peuvent être des puissances exactes; autrement dit: l'équation $x^p - y^q = 1$ dans laquelle les inconnues sont entières et positives, n'admèt qu'une seule solution.

In other words, Catalan proposed the following.

Conjecture *(E. Catalan, 1844) The only two consecutive numbers in the sequence of perfect powers of natural numbers*

$$1, 4, 8, 9, 16, 25, 27, 32, 36, 49, 64, 81, 100, 121, 125, 128, 144, 169, \ldots$$

are 8 and 9.

When $k \geq 2$ is fixed, the kth powers of natural numbers are necessarily far apart. However, when one varies k, two powers can be closer to one another than one might expect. For instance, we have $2^7 - 5^3 = 128 - 125 = 3$ and $13^3 - 3^7 = 2197 - 2187 = 10$. Catalan conjectured that the only powers for which the difference is as small as 1 are 3^2 and 2^3.

Phrased in yet another way, Catalan conjectured that for exponents $p, q \geq 2$, the Diophantine equation

$$x^p - y^q = 1$$

admits no solution in natural numbers other than the one given by $x = 3$, $p = 2$ and $y = 2$, $q = 3$.

R. Schoof, *Catalan's Conjecture*, DOI: 10.1007/978-1-84800-185-5_1,
© Springer-Verlag London Limited 2008

Apparently, Catalan himself did not get very far in solving the problem. We read this in a note [12] that was published more than forty years after his 1844 letter. Here Catalan reports on his early attempts:

> Après avoir perdu près d'une année à la recherche d'une démonstration qui fuyait toujours, j'abandonnerai cette recherche fatigante.

The *Société Belge des Professeurs de Mathématique d'Expression Française* has published a book on the academic and political activities of Eugène Catalan [20]. It contains a reproduction of a painting of Catalan which is at present in the possession of the Université de Liège (Fig. 1.1).

In this book, we mainly concentrate on Preda Mihăilescu's proof of Catalan's famous conjecture. We discuss earlier work only when it is relevant to our

Fig. 1.1 Eugène Catalan (1814–1894)

Fig. 1.2 Rob Tijdeman. (Reproduced by the kind permission of Rob Tijdeman.)

presentation of the proof. For an overview of earlier work on the conjecture, see [7, 14, 15, 35, 38]. We only mention one important result: In 1976, Rob Tijdeman [49] (Fig. 1.2) showed that there exist only *finitely many* pairs of consecutive perfect powers. His proof is based on the theory of linear forms in logarithms. Unfortunately, the bound on the size of the solutions that came out of Tijdeman's proof is astronomical. There remained a large gap between the relatively small exponents p, q for which Catalan's equation had been solved and Tijdeman's estimates. In the successive years, this gap was narrowed considerably by various people [2, 6, 17, 19, 26, 32, 33, 34, 44, 46]. But the gap remained very large. In 1999, refinements of Tijdeman's estimates were shown to imply Catalan's conjecture when both exponents p, q exceed $7.78 \cdot 10^{16}$. On the other hand, elaborate computer calculations had proven the conjecture when one of p, q is smaller than 10^5. See [34] for more information.

Between 2000 and 2003, Preda Mihăilescu (Fig. 1.3) proved three theorems concerning Catalan's conjecture:

Let p, q be odd primes and suppose that x, y are nonzero integers for which $x^p - y^q = 1$.

Theorem I *(P. Mihăilescu, 2000) We have*

$$p^{q-1} \equiv 1 \pmod{q^2} \quad \textit{and} \quad q^{p-1} \equiv 1 \pmod{p^2}.$$

Fig. 1.3 Preda Mihăilescu. (Reproduced by the kind permission of Robert Tichy.)

Theorem II *(P. Mihăilescu, 2002) We have*

$$p \equiv 1(\mathrm{mod}\ q) \quad or \quad q \equiv 1(\mathrm{mod}\ p).$$

Theorem III *(P. Mihăilescu, 2003) We have*

$$p < 4q^2 \quad and \quad q < 4p^2.$$

Mihăilescu's proofs of Theorems I, II, and III appear in [35], [36], and [37], respectively. We show now that these three theorems together lead to a proof of Catalan's conjecture. This proof does not rely on Tijdeman's work or on the computer calculations mentioned above.

Main theorem *The only solutions of the equation*

$$x^p - y^q = 1$$

in integers $p, q \geq 2$ and nonzero integers x, y are given by $p = 2$, $q = 3$ and $x = \pm 3$, $y = 2$.

Proof It is indeed true that $(\pm 3)^2 - 2^3$ is equal to 1. To prove that there are no other solutions, it suffices to show that there are no other solutions when the exponents p and q are *prime numbers*. See Exercise 1.1.

The case $q = 2$ was taken care of by V. Lebesgue [27] in 1850. The case $p = 2$ was dealt with by Ko Chao [21] in 1965. See chapters 2 and 3 for detailed proofs of these two results.

Therefore, we may assume that both p and q are odd prime numbers. We suppose that x, y are nonzero integers satisfying $x^p - y^q = 1$ and we derive a contradiction from this assumption. The problem is symmetric in p and q, as one sees by replacing (x, y, p, q) by $(-y, -x, q, p)$ in the equation $x^p - y^q = 1$.

By Theorem II, we have either $p \equiv 1 \pmod{q}$ or $q \equiv 1 \pmod{p}$. By symmetry, we may assume that $p \equiv 1 \pmod{q}$. It follows then from Theorem I and Exercise 1.2 that one even has $p \equiv 1 \pmod{q^2}$. Therefore, $p = 1 + kq^2$ for some integer $k \geq 1$. Theorem III implies then that $k \leq 3$. Since both $1 + q^2$ and $1 + 3q^2$ are even and hence cannot be prime, we must have $k = 2$. Since $p = 2q^2 + 1$ is divisible by 3 when $q \neq 3$, we must have $q = 3$. This implies $p = 19$. But this is impossible because we have $3^{18} \not\equiv 1 \pmod{19^2}$, contradicting Theorem I.

This proves the main theorem.

Theorem I is Mihăilescu's so-called double Wieferich criterion [35]. Proved in 2000, it shows that if exceptions to Catalan's conjecture exist, they are very rare indeed. But double Wieferich pairs do exist: $p = 83$, $q = 4871$ is an example. See Exercise 1.4 for other examples. Mihăilescu's main result [36] is Theorem II. In combination with certain estimates from the theory of linear forms in logarithms and a computer calculation, it led in 2002 to a complete proof of Catalan's conjecture. To reach this point, Yuri Bilu's (Fig. 1.4) efforts to understand Mihăilescu's original proof have been of great importance [3]. The proof that we present here avoids both the theory of linear forms in logarithms and the computer calculation. It replaces these by Mihăilescu's Theorem III, the proof of which involves only the arithmetic of cyclotomic fields.

The proofs of Theorems II and III that we give work only when p, q are not very small. To take care of small exponents, we use the following result. We present Mihăilescu's proof [37] of it in chapter 8. This result is much stronger than what we need to make the proofs of Theorems II and III work. We need it only for p or q less than or equal to 5.

Theorem IV *Let p, q be odd primes. Suppose that we have $p \leq 41$ or $q \leq 41$. Then Catalan's equation $x^p - y^q = 1$ admits no nonzero solutions $x, y \in \mathbf{Z}$.*

There are two main ingredients in the proofs of Theorems I–IV. The first one is Runge's method [40]. This method can be said to have already played a role in the proofs of certain earlier results regarding Catalan's conjecture. The second ingredient is the theory of cyclotomic fields. In particular, Stickelberger's theorem and Thaine's theorem play central roles in the proofs. As we already pointed out,

Fig. 1.4 Yuri Bilu. (Reproduced by the kind permission of Francine Delmer.)

Mihăilescu's proof does not make any use of Tijdeman's result [49] and does not rely on any computer calculation.

The first few chapters of this book regard relevant earlier work. In chapter 2 and 3, we discuss Catalan's equation $x^p - y^q = 1$ when $q = 2$ and $p = 2$, respectively. These results are very classical [21, 27]. Chapter 4 is devoted to the Diophantine equation $x^2 - y^3 = 1$, which Euler solved in 1738. We solve it following W. McCallum [31], who dealt with the problem as a student participating in a 1977 honors project at the University of New South Wales. We are then reduced to the case where both p and q are odd primes. In chapter 5, we discuss C. Runge's method to effectively bound the integral solutions of certain Diophantine equations.

In chapter 6, we use this method to prove J.W.S. Cassels' 1960 result [8], which says that when x,y are nonzero integers satisfying $x^p - y^q = 1$, then q divides x and p divides y. We show that Cassels' theorem easily implies that any nonzero solution $x, y \in \mathbf{Z}$ to Catalan's equation is necessarily very large with respect to the exponents p and q.

The remaining chapters regard Mihăilescu's proofs and involve the theory of cyclotomic fields. Our presentation of Mihăilescu's work is as follows [43]. To a nonzero solution of Catalan's equation $x^p - y^q = 1$ we associate the element $x - \zeta_p$ of a certain obstruction group H. Here ζ_p denotes a primitive pth root of unity and the group H is defined as

$$H = \left\{ \alpha \in \mathbf{Q}(\zeta_p)^* : \mathrm{ord}_{\mathfrak{r}}(\alpha) \equiv 0 (\mathrm{mod}\ q)\ \text{for all prime ideals } \mathfrak{r} \neq \mathfrak{p} \right\} / \mathbf{Q}(\zeta_p)^{*q},$$

where $\mathfrak{p}$ denotes the unique prime ideal of the ring $\mathbf{Z}[\zeta_p]$ lying over p. In chapter 7, we explain that the group H is finite and that there is a natural exact sequence of $\mathbf{F}_q[G]$-modules

$$0 \longrightarrow E_p/E_p^q \longrightarrow H \longrightarrow Cl_p[q] \longrightarrow 0,$$

where G is the Galois group of $\mathbf{Q}(\zeta_p)$ over $\mathbf{Q}$ and E_p and Cl_p denote the group of p-units and ideal class group, respectively, of the cyclotomic field $\mathbf{Q}(\zeta_p)$. See chapter 7 for more details.

On the one hand, the fact that $x - \zeta_p$ comes from a solution to Catalan's equation is shown to imply that the $\mathbf{F}_q[G]$-submodule of H generated by $x - \zeta_p$ is *large* in various senses. For instance, Cassels' result mentioned earlier can be viewed as saying that $x - \zeta_p$ is not contained in a certain index-q submodule of H. Theorem 8.3, Proposition 11.3, and Theorem 12.4 each state that the $\mathbf{F}_q[G]$-submodule of H generated by $x - \zeta_p$ is *large* in a certain sense.

On the other hand, the general theory of cyclotomic fields implies that the $\mathbf{F}_q[G]$-module generated by $x - \zeta_p$ is also *small* from various points of view. Indeed, Stickelberger's classical theorem and Thaine's 1988 theorem provide elements in the group ring $\mathbf{Z}[G]$ that *annihilate* the obstruction group H or at least certain parts of it. For instance, Mihăilescu's "double Wieferich criterion" is proved using Stickelberger's theorem. His result, which is our Theorem I, can be viewed as saying that $x - \zeta_p$ is contained in a proper submodule of the obstruction group H. More precisely, $x - \zeta_p$ is contained in the "Selmer group" S defined by

$$S = \{\alpha \in H : \alpha \text{ is a } \mathfrak{q}\text{-adic } q\text{th power for each prime } \mathfrak{q} \text{ lying over } q\}.$$

Corollary 10.3, Proposition 11.2, and Theorem 14.1 each state that the $\mathbf{F}_q[G]$ module generated by $x - \zeta_p$ is *small* in a certain sense.

Confronting both type of statements leads to contradictions. The conclusion is then that Catalan's equation $x^p - y^q = 1$ does not admit any nonzero solutions other than the solution $(\pm 3)^2 - 2^3 = 1$.

In chapter 8, we prove Theorem IV, while in chapter 9, we discuss some of the basic properties of the Stickelberger ideal. These are then used in chapters 10 and 11, where we prove Theorems I and III, respectively. The remaining chapters are devoted to the proof of Theorem II. The relevant Runge argument is given in chapter 12. In chapter 13, we discuss some elementary properties of semisimple group rings. In chapter 14, we exploit these in the proof of Theorem II. The proof involves Thaine's theorem, which we prove in chapter 16. Our proof of Thaine's theorem makes use of Chebotarëv's density theorem, which we prove in chapter 15.

Exercises

1.1 Show that it suffices to prove Catalan's conjecture for prime exponents: If the only solution to $x^p - y^q = 1$ in nonzero integers x, y and *prime numbers p,q* is the one given by $(\pm 3)^2 - 2^3 = 1$, then Catalan's conjecture is true.

1.2 Let q be prime and suppose $x \in \mathbf{Z}$ satisfies $x \equiv 1 (\bmod q)$. Show that $x^{q-1} \equiv 1 (\bmod q^2)$ implies that $x \equiv 1 (\bmod q^2)$.

1.3 Check that $3^{18} \not\equiv 1 (\bmod 19^2)$.

1.4 Show that the pairs of primes $(p, q) = (2, 1093)$ and $(911, 318917)$ are double Wieferich pairs, i.e., they satisfy the congruences of Theorem I.

1.5 A *proper* power is a natural number of the form n^k for some natural number n and some exponent $k \in \mathbf{Z}_{\geq 2}$.

   (a) Show that every $m \not\equiv 2 (\bmod 4)$ is the difference of two proper powers.
   (b) (Research problem) Investigate whether $m = 2, 6, 10, 14, 18, \ldots$ are differences of proper powers.

2
The Case "$q = 2$"

In this chapter, we deal with the case where the exponent q in Catalan's equation $x^p - y^q = 1$ is equal to 2. The proof is by a 2-adic argument [27]. It exploits the arithmetic of the ring of Gaussian integers $\mathbf{Z}[i]$. See Exercise 2.2.

Proposition 2.1 *(V.A. Lebesgue, 1850) For any exponent $p \geq 2$, the Diophantine equation*

$$x^p = y^2 + 1$$

has no solution in nonzero integers x, y.

Proof   Suppose $x, y \in \mathbf{Z}$ satisfy $x^p = y^2 + 1$. If p is even, we have $(x^{p/2} - y)$ $(x^{p/2} + y) = 1$ and hence $x^{p/2} - y = x^{p/2} + y = \pm 1$. Subtracting the equations, we find that $y = 0$. Since $y \neq 0$, we may therefore assume that p is odd.

Considering the equation modulo 4, we see that y is even and x is odd. See Exercise 2.1. It follows that $1 + iy$ and $1 - iy$ are coprime in the ring $\mathbf{Z}[i]$. By Exercise 2.2, the ring $\mathbf{Z}[i]$ is a unique factorization domain with precisely four units. Since p is odd, it follows that the units of $\mathbf{Z}[i]$ are pth powers. Since $(1+iy)(1-iy)$ is a pth power, Exercise 2.3 then implies that

$$1 + iy = c^p, \qquad \text{for some } c \in \mathbf{Z}[i].$$

Adding the complex conjugate equation, it follows that

$$2 = c^p + \bar{c}^p = (c + \bar{c})(c^{p-1} - c^{p-2}\bar{c} + \ldots + \bar{c}^{p-1}).$$

Therefore, the integer $c + \bar{c}$ divides 2. Since $c + \bar{c}$ is even, it is equal to ± 2. It follows that $c = \pm(1 + bi)$ for some $b \in \mathbf{Z}$. Moreover, since y is even, Exercise 2.2(d) implies that c is not divisible by $1 + i$ and that b is also even. We get

$$(1 + bi)^p + (1 - bi)^p = \pm 2.$$

R. Schoof, *Catalan's Conjecture*, DOI: 10.1007/978-1-84800-185-5_2,
© Springer-Verlag London Limited 2008

Reducing this equality modulo the ideal $8\mathbf{Z}[i]$, we see that we have the plus sign on the right. Using the binomial expansion, we find the following relation among binomial coefficients $\binom{p}{k}$:

$$\binom{p}{2}(bi)^2 + \binom{p}{4}(bi)^4 + \ldots + \binom{p}{p-1}(bi)^{p-1} = 0.$$

Suppose $b \neq 0$. See Exercise 2.5 for the basic properties of p-adic valuations. We claim that the 2-adic valuation of the first term is *strictly smaller* than the 2-adic valuation of each of the other terms. In other words, we claim

$$\mathrm{ord}_2\left(\binom{p}{k}(bi)^k\right) > \mathrm{ord}_2\left(\binom{p}{2}(bi)^2\right),$$

for all even k satisfying $4 \leq k \leq p-1$. This follows from the fact that the 2-adic valuation of

$$\binom{p}{k}\binom{p}{2}^{-1}(bi)^{k-2} = \binom{p-2}{k-2}\frac{2}{k(k-1)}(bi)^{k-2}$$

is strictly positive. Indeed, by Exercise 2.4, the binomial coefficient $\binom{p-2}{k-2}$ is an integer and, by Exercise 2.6, we have for all even $k \geq 4$,

$$\mathrm{ord}_2(2(bi)^{k-2}) \geq k-1 > \frac{\log k}{\log 2} \geq \mathrm{ord}_2(k) = \mathrm{ord}_2(k(k-1)).$$

A repeated application of Exercise 2.5 now implies that $\mathrm{ord}_2(\binom{p}{2}(bi)^2)$ is equal to the 2-adic valuation of the entire sum, which is zero. This is only possible when $b = 0$. But then we have $c = 1$ and hence $y = 0$. This proves the proposition.

Exercises

2.1 Let $x \in \mathbf{Z}$. Show that x^2 is congruent to 0 or 1 (mod 4). Show that x^2+2 cannot be a square modulo 4.

2.2 Let $\mathbf{Z}[i]$ be the subring of $\mathbf{C}$ defined by $\mathbf{Z}[i] = \{a + bi : a, b \in \mathbf{Z}\}$.

   (a) Show that $\mathbf{Z}[i]$ is a principal ideal domain and hence a unique factorization domain.

   (b) Show that the unit group $\mathbf{Z}[i]^*$ is equal to $\{\pm 1, \pm i\}$.

   (c) Show that the $\mathbf{Z}[i]$-ideal generated by $1+i$ is prime and that its square is generated by 2.

   (d) Show that $a + bi$ is divisible by $1+i$ if and only if $a \equiv b \mod 2$.

2.3 Let $n \geq 1$ be an integer.

(a) Let R be a unique factorization domain. Suppose that a, b are coprime elements of R with the property that $ab = c^n$ for some nonzero $c \in R$. Show that there are $r, s \in R$ and a unit $u \in R^*$ for which $a = ur^n$ and $b = u^{-1}s^n$.

(b) Take $R = \mathbf{Z}$ and n odd in part (a). Show that $a = r^n$ and $b = s^n$ for certain $r, s \in \mathbf{Z}$.

(c) Do the same for $R = \mathbf{Z}[i]$.

2.4 Let $n \in \mathbf{Z}$.

(a) Show that for any $n \in \mathbf{Z}_{>0}$ and any natural number k, the binomial coefficient

$$\binom{n}{k} = \frac{n \cdot (n-1) \cdots (n-k+1)}{k \cdot (k-1) \cdots 1}$$

is an integer.

(b) Show that for $n > 0$, we have $\sum_{k=0}^{n} \binom{n}{k} = 2^n$ and hence $\binom{n}{k} \leq 2^n$ for all $0 \leq k \leq n$.

2.5 Let p be a prime. For any nonzero $x \in \mathbf{Z}$, let $\mathrm{ord}_p(x)$ denote the *p-adic valuation* of x. In other words, $\mathrm{ord}_p(x)$ is the number of factors p that occur in the prime factorization of x. For $x \in \mathbf{Q}^*$, we write $x = y/z$ for some $y, z \in \mathbf{Z}$ and set $\mathrm{ord}_p(x) = \mathrm{ord}_p(y) - \mathrm{ord}_p(z)$. It is convenient to set $\mathrm{ord}_p(0) = +\infty$. For $x, y \in \mathbf{Q}$, show that $\mathrm{ord}_p(xy) = \mathrm{ord}_p(x) + \mathrm{ord}_p(y)$. Show that $\mathrm{ord}_p(x + y) \geq \min(\mathrm{ord}_p(x), \mathrm{ord}_p(y))$, with equality holding when $\mathrm{ord}_p(x)$ and $\mathrm{ord}_p(y)$ are distinct.

2.6 Let p be a prime and let n be a natural number. Show that $\mathrm{ord}_p(n) \leq \log n / \log p$, with equality holding when n is a power of p.

3
The Case "$p = 2$"

In this chapter, we deal with the case where the exponent p in Catalan's equation $x^p - y^q = 1$ is equal to 2. This was first done in 1965 by Ko Chao [21]. The proof we present here is due to E.Z. Chein [13]. In chapter 5, we give an alternative proof of Lemma 3.2, based on Runge's method.

Lemma 3.1 *Let $q \geq 3$ be an odd integer and suppose that x, y are nonzero integers satisfying $x^2 - y^q = 1$. Then*

(i) replacing x by $-x$ if necessary, we have

$$\begin{cases} x - 1 = 2^{q-1}a^q, \\ x + 1 = 2b^q, \\ \quad y = 2ab, \end{cases}$$

for coprime integers $a, b \in \mathbf{Z}$ satisfying $\gcd(2a, b) = 1$;
(ii) we have $y \geq 2^{q-1} - 2$.

Proof (i) We have $(x - 1)(x + 1) = y^q$. If x is even, then the factors $x \pm 1$ of $x^2 - 1$ are coprime and, by Exercise 2.3, both $x - 1$ and $x + 1$ are qth powers. Since these qth powers differ only by 2, they are equal to ± 1 and we have $x = 0$, which is not the case. Therefore, x is odd and hence y is even. It follows that 2^q divides $(x - 1)(x + 1)$. Changing the sign of x if necessary, we may assume that $x \equiv 1 \pmod 4$. This implies that $(x + 1)/2$ is odd and hence that 2^{q-1} divides $x - 1$. Since $\gcd(x - 1, x + 1)$ is equal to 2, it follows that the two factors on the left-hand side of the equality

$$\left(\frac{x-1}{2^{q-1}} \right) \left(\frac{x+1}{2} \right) = \left(\frac{y}{2} \right)^q$$

are coprime integers. By Exercise 2.3, both $(x - 1)/2^{q-1}$ and $(x + 1)/2$ are qth powers: $(x - 1)/2^{q-1} = a^q$ and $(x + 1)/2 = b^q$ for certain $a, b \in \mathbf{Z}$. Moreover, b is odd and we have $\gcd(a, b) = 1$. This proves (i).

R. Schoof, *Catalan's Conjecture*, DOI: 10.1007/978-1-84800-185-5_3,
© Springer-Verlag London Limited 2008

To prove *(ii)*, we subtract the first two equations of part *(i)* from one another. We find $2b^q \equiv 2 \pmod{2^{q-1}}$ and hence $b^q \equiv 1 \pmod{2^{q-2}}$. Since q is odd and the order of the group $(\mathbf{Z}/2^{q-2}\mathbf{Z})^*$ is a power of 2, this implies $b \equiv 1 \pmod{2^{q-2}}$. Since $b \neq 1$, we have therefore that $|b| \geq 2^{q-2} - 1$. From the fact that y is positive, it follows then that $y = 2ab \geq 2^{q-1} - 2$, as required.

The inequality of part *(ii)* is used in chapter 5 and plays no role here. The following two lemmas imply Ko Chao's theorem.

Lemma 3.2 *Let $q \geq 3$ be prime and let x, y be nonzero integers satisfying $x^2 - y^q = 1$. Then we have $x \equiv 0 \pmod{q}$.*

Proof Suppose that we have $x^2 - y^q = 1$ but $x \not\equiv 0 \pmod{q}$. We have

$$(y+1)\left(\frac{y^{q+1}}{y+1}\right) = x^2.$$

By Exercise 3.2, the gcd of the factors $y + 1$ and $(y^q + 1)/(y + 1)$ divides q. Since $x \not\equiv 0 \pmod{q}$, the gcd is therefore equal to 1. Exercise 2.3 implies then that each of the factors is a square times ± 1. Since $y^q + 1$ is a square, we have $y \geq 0$. Therefore, both $y + 1$ and $(y^q + 1)/(y + 1)$ are positive. It follows that

$$y + 1 = u^2 \qquad \text{for some nonzero } u \in \mathbf{Z}.$$

Therefore, both $(X, Y) = (x, y^{(q-1)/2})$ and $(X, Y) = (u, 1)$ are solutions to the equation

$$X^2 - y \cdot Y^2 = 1.$$

Since $y = u^2 - 1$ is not zero, it must be positive. By Exercise 3.4, the number y is not a square, and so we are dealing with a *Pell* equation. The ring $\mathbf{Z}[\sqrt{y}]$ is a subring of the ring of integers of a real quadratic number field and $x + y^{(q-1)/2}\sqrt{y}$ is a unit of $\mathbf{Z}[\sqrt{y}]$. By Exercise 3.4, the unit group of $\mathbf{Z}[\sqrt{y}]$ is generated by -1 and $u + \sqrt{y}$. It follows that

$$x + y^{(q-1)/2}\sqrt{y} = \pm(u + \sqrt{y})^m \qquad \text{for a certain } m \in \mathbf{Z}.$$

Since $(u + \sqrt{y})^{-1} = -(-u + \sqrt{y})$, we may change the sign of u and assume that $m \geq 0$. Reducing the equation modulo the ideal $y\mathbf{Z}[\sqrt{y}]$, we find

$$x \equiv \pm(u^m + mu^{m-1}\sqrt{y}) \pmod{y\mathbf{Z}[\sqrt{y}]}.$$

Since the elements $1, \sqrt{y}$ form a **Z**-basis for the additive group of $\mathbf{Z}[\sqrt{y}]$, it follows that $mu^{m-1} \equiv 0 \pmod{y}$. This implies that y divides mu^{m-1}. Since

$y + 1 = u^2$, Exercise 2.1 implies that y is even and u is odd. It follows that m is even. Therefore, we have

$$x + y^{(q-1)/2}\sqrt{y} = \pm(u^2 + y + 2u\sqrt{y})^{m/2}.$$

Reducing this equation modulo the ideal $u\mathbf{Z}[\sqrt{y}]$, we find

$$x + y^{(q-1)/2}\sqrt{y} \equiv \pm y^{m/2} \pmod{u\mathbf{Z}[\sqrt{y}]}.$$

Inspection of the $\sqrt{y}$-coefficients shows that u divides $y^{(q-1)/2}$. Since $y + 1 = u^2$, we have $\gcd(u, y) = 1$. It follows that $u = \pm1$ and hence $y = 0$.

This contradiction shows that $x \equiv 0 \pmod{q}$, as required.

Lemma 3.3 *Let $q \geq 3$ be prime and let x, y be nonzero integers satisfying $x^2 - y^q = 1$. Then we have $x \equiv \pm3 \pmod{q}$.*

Proof In view of the previous lemma, we may assume $q \geq 5$. Changing the sign of x if necessary, Lemma 3.1 implies $x - 1 = 2^{q-1}a^q$ and $x + 1 = 2b^q$ for certain $a, b \in \mathbf{Z}$ with $\gcd(2a, b) = 1$. It follows that we have

$$b^{2q} - (2a)^q = \left(\frac{x+1}{2}\right)^2 - 2(x-1) = \left(\frac{x-3}{2}\right)^2.$$

Factoring the left-hand side, we find

$$(b^2 - 2a)\left(\frac{b^{2q} - (2a)^q}{b^2 - 2a}\right) = \left(\frac{x-3}{2}\right)^2.$$

Since $2a$ and b^2 are coprime, Exercise 3.2 implies that the gcd of the two factors divides q. We claim that the gcd is actually equal to q. Indeed, if it is 1, Exercise 2.3 implies that both factors are equal to a square times ±1. Since $b^{2q} - (2a)^q$ is a square, we have $b^{2q} \geq (2a)^q$ and hence $b^2 \geq 2a$. It follows that we have the plus sign and that $b^2 - 2a$ is a *square*. Since y is not zero, neither is a. This means that $b^2 - 2a = c^2$ for some square c^2 that is different from b^2. The squares nearest to b^2 are $(b \pm 1)^2$. Since $|(b \pm 1)^2 - b^2| \geq 2|b| - 1$, we have $|c^2 - b^2| = 2|a| \geq 2|b| - 1$ and hence $|a| \geq |b|$. However, from the inequalities $q \geq 5$ and $|x| \geq 2$, it follows that

$$|a|^q = \frac{|x-1|}{2^{q-1}} \leq \frac{|x-1|}{16} < \frac{|x+1|}{2} = |b|^q,$$

showing that $|a| < |b|$.

This contradiction shows that the gcd of $b^2 - 2a$ and the cofactor $(b^{2q} - (2a)^q)/(b^2 - 2a)$ cannot be 1 and therefore must be equal to q. As a consequence, $\left(\frac{x-3}{2}\right)^2$ is divisible by q so that $x \equiv 3 \pmod{q}$, as required.

Corollary 3.4 *(Ko Chao, 1965) Let $q \geq 5$ be prime. Then the Diophantine equation*

$$x^2 - y^q = 1$$

has no nonzero solutions $x, y \in \mathbf{Z}$.

Proof Suppose that $x, y \in \mathbf{Z}$ is a nonzero solution. By Lemmas 3.2 and 3.3, we have $x \equiv 0 \pmod{q}$ as well as $x \equiv \pm 3 \pmod{q}$. Since $q \neq 3$, this is clearly impossible. Therefore, there are no solutions and the result follows.

Exercises

3.1 Let n be an odd integer. Show that $\frac{x^n - y^n}{x - y}$ is positive for any two distinct real numbers x, y.

3.2 Let q be a prime number and let x, y be distinct integers with $\gcd(x, y) = 1$. Show that

$$\gcd\left(\frac{x^q - y^q}{x - y}, x - y\right) \quad \text{divides} \quad q.$$

3.3 Let n be a natural number and let x, y be distinct integers that have the same sign. Suppose that $|x| < |y|$. Show that $|y^n - x^n| \geq n|x|^{n-1}$. Show that the inequality is strict when $n \geq 2$.

3.4 Let $u \in \mathbf{Z}_{>1}$ and let $y = u^2 - 1$.

    (a) Show that y is not a square.

    (b) Show that -1 and $\varepsilon = u + \sqrt{y}$ generate the unit group of the ring $\mathbf{Z}[\sqrt{y}]$. [Hint: Let $j : \mathbf{Z}[\sqrt{y}] \hookrightarrow \mathbf{R}$ denote the embedding that maps $\sqrt{y}$ to the positive square root of y in $\mathbf{R}$. Let η be an arbitrary unit of $\mathbf{Z}[\sqrt{y}]$. By multiplying by $\pm\varepsilon^k$ for some $k \in \mathbf{Z}$, we may assume that $1 \leq j(\eta) < j(\varepsilon)$. We claim that $\eta = 1$. To see this, suppose that $\eta = a + b\sqrt{y}$ for two integers $a, b \geq 0$. If $\eta \neq 1$, we have $b \geq 1$. Since the norm of η is equal to ± 1, this implies that $a^2 = \pm 1 + b^2 y \geq y - 1 = u^2 - 2$. Therefore, $a \geq \sqrt{u^2 - 2} > u - 1$. It follows that $a \geq u$ and hence $j(\eta) \geq j(\varepsilon)$, a contradiction.]

4
The Nontrivial Solution

In this chapter we deal with the exponents $p = 2$ and $q = 3$. These were not discussed in Chapter 3. In this case, Catalan's equation $x^2 = y^3 + 1$ describes an elliptic curve E defined over $\mathbf{Q}$. Standard "descent" methods [52, App. IV, p. 140] show that the group $E(\mathbf{Q})$ of rational points is finite and of order 6. Indeed, $E(\mathbf{Q})$ consists of the points $(x, y) = (0, -1), (\pm 1, 0), (\pm 3, 2)$, and the point at infinity. It follows that the only integral solutions of the equation $x^2 - y^3 = 1$ with $x, y \neq 0$ are given by $(x, y) = (\pm 3, 2)$, as required. Euler's proof [16], presented in the Appendix, can be interpreted as such a proof by descent. See [52, XVI].

In this chapter, we present a different solution. It is due to W. McCallum [31]. He reduces the problem to a Thue equation, which he solves by means of Skolem's method. See [9, s. 10.10] for this method.

McCallum's method involves arithmetic in the ring $\mathbf{Z}[\sqrt[3]{2}]$. We introduce some terminology. For every element x of the number field $\mathbf{Q}(\sqrt[3]{2})$, there are unique $a, b, c \in \mathbf{Q}$ for which we have $x = a + b\sqrt[3]{2} + c\sqrt[3]{4}$. We call the coefficient c the $\sqrt[3]{4}$-*coefficient* of x.

Proposition 4.1 *Let η be the unit $\sqrt[3]{2} - 1$ of the ring $\mathbf{Z}[\sqrt[3]{2}]$. The only exponents $n \in \mathbf{Z}$ for which the $\sqrt[3]{4}$-coefficient of η^n is zero are $n = 0$ and 1.*

Proof The multiplicative inverse of η is easily checked to be $1 + \sqrt[3]{2} + \sqrt[3]{4}$. For every $m > 0$, the coefficients of $\left(1 + \sqrt[3]{2} + \sqrt[3]{4}\right)^m$ with respect to the basis $\{1, \sqrt[3]{2}, \sqrt[3]{4}\}$ are all positive. This follows easily by induction. Therefore, the $\sqrt[3]{4}$-coefficient of η^n is never zero when n is negative.

From now on we suppose that $n \geq 0$. Set $\pi = 1 + \sqrt[3]{2}$. This number generates the unique prime ideal lying over 3 in the ring $\mathbf{Z}[\sqrt[3]{2}]$. See Exercise 4.3. The proof is by a π-adic argument. For every integer $k \geq 0$, there are unique $a_k, b_k, c_k \in \mathbf{Q}$ such that $\pi^k = a_k + b_k\sqrt[3]{2} + c_k\sqrt[3]{4}$. The coefficients a_k, b_k, c_k are actually in $\mathbf{Z}$. For $k \leq 6$, they are listed in Table 4.1.

Since π^3 is equal to $\left(1 + \sqrt[3]{2}\right)^3 = 3 + 3\sqrt[3]{2} + 3\sqrt[3]{4}$, it is contained in the ideal $3\mathbf{Z}[\sqrt[3]{2}]$. Therefore, for every $k \geq 0$, the powers π^k are contained in the

R. Schoof, *Catalan's Conjecture*, DOI: 10.1007/978-1-84800-185-5_4,
© Springer-Verlag London Limited 2008

Table 4.1 Coefficients a_k, b_k, and c_k

k	0	1	2	3	4	5	6
a_k	1	1	1	3	9	21	45
b_k	0	1	2	3	6	15	36
c_k	0	0	1	3	6	12	27

ideal $3^{[\frac{k}{3}]}\mathbf{Z}[\sqrt[3]{2}]$. This implies that for every $k \geq 0$, the coefficients a_k, b_k, and c_k are divisible by $3^{[\frac{k}{3}]}$. Here $\left[\frac{k}{3}\right]$ denotes the integral part of $\frac{k}{3}$.

It is convenient to consider the number $-\frac{\eta}{2} = 1 - \frac{\pi}{2}$ rather than the unit η itself. We have

$$\left(-\frac{\eta}{2}\right)^n = \left(1 - \frac{\pi}{2}\right)^n = \sum_{k=0}^{n} \binom{n}{k} \left(-\frac{\pi}{2}\right)^k.$$

Suppose that for some $n \geq 2$, the $\sqrt[3]{4}$-coefficient of η^n vanishes. Then the same is true for the $\sqrt[3]{4}$-coefficient of $\left(-\frac{\eta}{2}\right)^n$. The $\sqrt[3]{4}$-coefficient of $\left(-\frac{\eta}{2}\right)^n$ is equal to the sum of the series of the $\sqrt[3]{4}$-coefficients of the terms of the sum above. Since the $\sqrt[3]{4}$-coefficients of the terms are equal to $\binom{n}{k} \left(-\frac{1}{2}\right)^k c_k$, we have

$$0 = \sum_{k=0}^{n} \binom{n}{k} \left(-\frac{1}{2}\right)^k c_k.$$

Since $c_0 = c_1 = 0$, the first two terms of the series vanish. By assumption, we have $n \neq 0$ or 1, so that we can divide by $n(n-1)$. Since we have

$$\frac{1}{n(n-1)} \binom{n}{k} = \frac{1}{k(k-1)} \binom{n-2}{k-2},$$

we find

$$0 = \sum_{k=2}^{n} \frac{1}{k(k-1)} \binom{n-2}{k-2} \left(-\frac{1}{2}\right)^k c_k.$$

By Exercise 2.4, the binomial coefficients are all integers. The coefficients c_k are divisible by $3^{[\frac{k}{3}]}$. Exercise 4.1 implies that for $k \geq 2$, the 3-adic valuation of $k(k-1)$ is at most $[k/3]$. It follows that all terms are rational numbers with denominators that are prime to 3. The inequality $\mathrm{ord}_3(k(k-1)) \leq [k/3]$ is strict when $k \geq 5$. It follows that the terms with $k \geq 5$ are congruent to 0 (mod 3). As one can see in Table 4.1, the coefficients c_2, c_3, and c_4 are equal to 1, 3, and 6, respectively.

Therefore, evaluating the sum modulo 3 and keeping in mind that $\frac{1}{2} \equiv 2$ (mod 3), we find

$$0 \equiv \frac{1}{2}\left(-\frac{1}{2}\right)^2 + \frac{1}{2\cdot 3}(n-2)\left(-\frac{1}{2}\right)^3 3 + \frac{1}{3\cdot 4}\frac{(n-2)(n-3)}{2}\left(-\frac{1}{2}\right)^4 6 \text{ (mod 3)}$$
$$\equiv -1 - (n+1) + (n+1)n \text{ (mod 3)}$$
$$\equiv n^2 + 1 \text{ (mod 3)}.$$

This is impossible, and the proposition follows.

Proposition 4.2 *The only nonzero solutions $x, y \in \mathbf{Z}$ to the Diophantine equation*

$$x^2 - y^3 = 1$$

are given by $(x, y) = (\pm 3, 2)$. In particular, the only nontrivial solutions $x, y \in \mathbf{Z}$ to Catalan's equation with exponents $p = 2$ and $q = 3$ are given by $(x, y) = (\pm 3, 2)$.

Proof Let $x, y \in \mathbf{Z}$ be a nonzero solution to the equation $x^2 - y^3 = 1$. We have

$$(x - 1)(x + 1) = y^3.$$

If x is even, the factors on the left are coprime and Exercise 2.3 implies that $x - 1 = r^3$ and $x + 1 = s^3$ for certain $u, v \in \mathbf{Z}$. It follows that $s^3 - r^3 = 2$. Exercise 3.3 shows that s and r cannot have the same sign. It follows easily that $s = -r = 1$. But then we have $x = 0$, which is not the case. Therefore, x is odd. Changing the sign of x if necessary, we may assume that $x \equiv 1$ (mod 4). We have

$$\frac{x - 1}{4}\frac{x + 1}{2} = \left(\frac{y}{2}\right)^3.$$

The two factors on the left are coprime integers. Therefore, Exercise 2.3 implies that

$$\frac{x - 1}{4} = u^3 \quad \text{and} \quad \frac{x + 1}{2} = v^3, \quad \text{for certain } u, v \in \mathbf{Z}.$$

This shows that

$$v^3 - 2u^3 = 1.$$

In other words, the number $\varepsilon = v - u\sqrt[3]{2}$ is a unit of the ring $\mathbf{Z}[\sqrt[3]{2}]$. Exercise 4.4 implies that

$$\varepsilon = \pm\eta^n,$$

where $\eta = \sqrt[3]{2} - 1$. To see that we have the plus sign, we take norms. By Exercise 4.2, the norm from $\mathbf{Q}(\sqrt[3]{2})$ to $\mathbf{Q}$ of $\varepsilon = v - u\sqrt[3]{2}$ is equal to $v^3 - 2u^3 = 1$. On the other hand, the norm of η is 1. Therefore, we have $\varepsilon = +\eta^n$. Since the $\sqrt[3]{4}$-coefficient of ε vanishes, Proposition 4.1 implies that $n = 0$ or $n = 1$. In other words, we have

$$v - u\sqrt[3]{2} = 1 \quad \text{or} \quad v - u\sqrt[3]{2} = \sqrt[3]{2} - 1.$$

In the first case, we have $u = 0$, $v = 1$ and hence $y = 0$, which is not the case. In the second case, we have $u = v = -1$ and hence $x = -3$, $y = 2$.

This proves the proposition.

We see, incidentally, that the nonzero solution $(x, y) = (\pm 3, 2)$ satisfies $x \equiv 0 \pmod 3$, as predicted by Lemmas 3.2 and 3.3 of the previous chapter. However, these lemmas were never used in the proof of Proposition 4.2.

Exercises

4.1 Compute $\mathrm{ord}_3(k(k-1))$ for $2 \le k \le 10$. Show that the 3-adic valuation of $k(k-1)$ is at most $\left[\frac{k}{3}\right]$ for $k \ge 2$. Show that the inequality is strict when $k \ge 5$.

4.2 Let $f \in \mathbf{Q}[X]$ be an irreducible monic polynomial and let $\alpha \in \overline{\mathbf{Q}}$ denote a zero of f. Put $F = \mathbf{Q}(\alpha)$. Show that for every $t \in \mathbf{Q}$, the norm (from F to $\mathbf{Q}$) of $t - \alpha$ is equal to $f(t)$.

4.3 (a) Show that the ideal generated by 2 in the ring $\mathbf{Z}[\sqrt[3]{2}]$ is the cube of the ideal generated by $\sqrt[3]{2}$. Compute the residue field of the prime $\sqrt[3]{2}$.

(b) Show that the ideal generated by 3 is the cube of the ideal generated by $\pi = 1 + \sqrt[3]{2}$. Compute the residue field of the prime π.

4.4 The goal of this exercise is to show that the unit group of the ring $\mathbf{Z}[\sqrt[3]{2}]$ is generated by -1 and $\eta = \sqrt[3]{2} - 1$. Consider the injective ring homomorphism $\mathbf{Q}(\sqrt[3]{2}) \hookrightarrow \mathbf{R} \times \mathbf{C}$ determined by

$$\sqrt[3]{2} \mapsto \left(\sqrt[3]{2}, e^{2\pi i/3}\sqrt[3]{2}\right).$$

We denote it by $x \mapsto (x', x'')$. For $x \in \mathbf{Q}(\sqrt[3]{2})$, we define its "length" $\|x\|$ by $\|x\|^2 = |x'|^2 + 2|x''|^2$.

(a) Let $a, b, c \in \mathbf{Q}$ and put $x = a + b\sqrt[3]{2} + c\sqrt[3]{4}$. Show that

$$\|x\|^2 = 3\left(a^2 + b^2\sqrt[3]{4} + 2c^2\sqrt[3]{2}\right).$$

(b) Show that η' is a real number satisfying $\frac{1}{4} < \eta' < 1$. Let u be an arbitrary unit of $\mathbf{Z}[\sqrt[3]{2}]$. Show that for some integer k, either the unit $\varepsilon = \pm u\eta^k$ or $\varepsilon = \pm u^{-1}\eta^k$ satisfies $\frac{1}{2} < \varepsilon' < 1$.

(c) Let ε be a unit of $\mathbf{Z}[\sqrt[3]{2}]$ satisfying $\frac{1}{2} < \varepsilon' < 1$. Show that $|\varepsilon''|^2 < 2$ and hence $\|\varepsilon\|^2 < 5$. Use part (a) to show that $\varepsilon = \pm 1$.

5
Runge's Method

In this chapter, we describe C. Runge's method to determine the integer solutions of certain Diophantine equations (Fig. 5.1). Runge described his method in a 1887 paper in Crelle's journal [40]. In this section, we explain it along the lines of a short unpublished note by Yuri Bilu. This involves somewhat more mathematics than is required for the rest of the notes. The method is applied to prove theorems of Cassels and Mihăilescu that regard Catalan's equation.

Our explanation of Runge's method involves the theory of algebraic curves. It is certainly not indispensable to view Runge's procedure as the analysis of the poles and zeroes of certain functions on algebraic curves, as we now proceed to do. Indeed, there is no trace of this point of view in Cassels' paper [8]. Mihăilescu apparently only heard of Runge's method months after he proved Theorem 12.4. However, several mathematicians—the author included—have the feeling they understand better what is happening when they see the estimates and inequalities embedded in this kind of mathematical context. In addition, our geometric point of view seems to be close to the way Runge [40] himself looked at the matter.

It is *not necessary* to understand the material in this section in order to read the rest of this book. Many readers may find this section useful, but others may prefer to skip it. To accommodate the latter, the proofs in chapters 6 and 12 are written without any reference to this chapter. They consist of a series of estimates and inequalities and can be read for their own sake. Only at the end of each proof is the relationship to the point of view presented in this chapter explained.

Let C be a smooth, projective, absolutely irreducible algebraic curve over $\mathbf{Q}$. The complex points of C form a compact Riemann surface $C(\mathbf{C})$. It has a natural topology induced by the usual topology of $\mathbf{C}$. Let Y be a nonconstant element of the function field $\mathbf{Q}(C)$. In other words, Y is a nonconstant function on C and is defined over $\mathbf{Q}$. A point P in the set of $\mathbf{Q}$-rational points $C(\mathbf{Q})$ is called *integral* if the value $Y(P)$ of the function Y in the point P is in $\mathbf{Z}$. This integrality notion depends, of course, on the choice of the function Y. Let $Q \in C(\mathbf{C})$ be a pole of Y. It has coordinates in $\overline{\mathbf{Q}}$. Since Y is defined over $\mathbf{Q}$, the $\mathrm{Gal}(\overline{\mathbf{Q}}/\mathbf{Q})$-conjugates of Q are also poles of Y.

Proposition 5.1 *Using the notation above, let $U \subset C(\mathbf{C})$ be an open neighborhood of Q and its Galois conjugates. Then there are only finitely many integral points in $C(\mathbf{Q})$ that are not contained in U.*

R. Schoof, *Catalan's Conjecture*, DOI: 10.1007/978-1-84800-185-5_5,
© Springer-Verlag London Limited 2008

Fig. 5.1 Carl Runge (1856–1927)

Proof By the Riemann–Roch theorem, there exists a nonconstant function in $\overline{\mathbf{Q}}(C)$ that has no poles outside Q. Multiplying this function by its $\mathrm{Gal}(\overline{\mathbf{Q}}/\mathbf{Q})$-conjugates, we obtain a nonconstant function Z in the function field $\mathbf{Q}(C)$ that has no poles outside the set $Q = Q_1, Q_2, \ldots, Q_t$ of Galois conjugates of Q. The function field $\mathbf{Q}(C)$ has finite degree over the field $\mathbf{Q}(Y)$. By Exercise 5.3, the function Z is *integral* over the ring $\mathbf{Q}[Y]$. By multiplying Z by a suitable nonzero integer, it becomes even integral over $\mathbf{Z}[Y]$. As a consequence, for any integral point $P \in C(\mathbf{Q})$, the rational number $Z(P)$ is integral over $\mathbf{Z}$. Gauss's lemma implies then that $Z(P)$ is an ordinary integer.

Now, since the points $Q_1, Q_2, \ldots, Q_t$ are the *only* poles of the function Z, the continuous function $P \mapsto |Z(P)|$ from the set $C(\mathbf{C}) - \{Q_1, Q_2, \ldots, Q_t\}$ to $\mathbf{R}$ is *bounded* on the compact set $C(\mathbf{C}) - U$. Therefore, there are only finitely many possibilities for the integer $Z(P)$. Since the function Z is not constant, there are only finitely many points $P \in C(\mathbf{Q})$ for which $Z(P)$ has a given value. It follows that there are only finitely many integral points on C that are not in U, as required.

In certain situations, one can in this way actually prove that the set of *all* integral points $P \in C(\mathbf{Q})$ is finite. This happens, for instance, if the pole divisor of the function Y has at least *two* $\mathrm{Gal}(\overline{\mathbf{Q}}/\mathbf{Q})$-orbits. Indeed, if Q and R are two nonconjugate poles of Y, we can find open neighborhoods U of Q and its conjugates and V of R and its conjugates with the property that $U \cap V = \emptyset$. Then we have

$$C(\mathbf{Q}) \subset (C(\mathbf{C}) - U) \cup (C(\mathbf{C}) - V),$$

and an application of Proposition 5.1 to the open sets U and V shows that the set of integral points in $C(\mathbf{Q})$ is finite.

Another favorable situation occurs when all conjugates of the pole Q of Y are complex rather than real. In this case, there exists an open neighborhood U of Q and its conjugates with the property that $U \cap C(\mathbf{R}) = \emptyset$. It follows that all rational points $P \in C(\mathbf{Q})$ are not contained in U. Therefore, $|Z(P)|$ can be bounded for all integral points $P \in C(\mathbf{Q})$.

The existence of the function Z is guaranteed by the Riemann–Roch theorem, but in practice one explicitly writes down a function Z that visibly has no poles outside the pole divisor of Y. This is done by showing that $|Z(P)|$ remains bounded as P approaches any of the poles of Y that are not conjugate to Q. The very existence of these bounds implies then that the function Z has no poles outside Q and its conjugates.

If the bound on the absolute values of $Z(P)$ for $P \notin U$ is explicit, then the approach described here leads to an effective procedure to determine all integral points $P \in C(\mathbf{Q})$ that are not in U.

Before illustrating the method by means of some explicit examples, we recall a basic fact from analysis.

Lemma 5.2 *Let $c > 0$ and let $F(t)$ be a function of a real variable that is $m + 1$ times differentiable in the interval $\{t \in \mathbf{R} : |t| < c\}$. Let $F_m(t)$ be the sum of the terms of degree $\leq m$ of the Taylor series around 0 of $F(t)$. Then for all t with $|t| < c$, there exists t' in the closed interval with endpoints 0 and t such that*

$$F(t) - F_m(t) = \frac{t^{m+1}}{(m+1)!} F^{(m+1)}(t').$$

Here $F^{(m+1)}$ denotes the $(m+1)$st derivative of the function F.

Proof This is just the the remainder term of the Taylor series expansion of $F(t)$ around 0. See [39, Thm. 5.15].

In the first example, we make use of the following estimate.

Lemma 5.3 *Let $F(t)$ be the function of a real variable t given by $F(t) = (1 + t + t^4)^{1/2}$. Let $F_2(t) = 1 + \frac{1}{2}t - \frac{1}{8}t^2$ denote the sum of the terms of degree at most 2 of the Taylor series of $F(t)$ around 0. Then we have*

$$|F(t) - F_2(t)| \leq \frac{1}{\sqrt{2}}|t|^3 \quad for \ -1/2 \leq t \leq 1/2.$$

Proof The polynomial $F_2(t)$ is also the sum of the terms of degree at most 2 of the Taylor series of the function $(1 + t)^{1/2}$ around 0. By Lemma 5.2, we have

$$|(1 + t)^{1/2} - F_2(t)| \leq \frac{|t|^3}{3!}\frac{3}{8}|1 + t'|^{-5/2},$$

for some t' between 0 and t. Since $|t'| \leq \frac{1}{2}$, the right-hand side is at most $\frac{1}{2\sqrt{2}}|t|^3$.

The difference between $F(t)$ and the function $(1 + t)^{1/2}$ is equal to

$$|F(t) - (1 + t)^{1/2}| = \frac{|t|^4}{(1 + t)^{1/2} + (1 + t + t^4)^{1/2}}.$$

Since $|t| \leq \frac{1}{2}$, this is also at most $\frac{1}{2\sqrt{2}}|t|^3$. Combining the two estimates, we get

$$|F(t) - F_2(t)| \leq |F(t) - (1 + t)^{1/2}| + |(1 + t)^{1/2} - F_2(t)|$$
$$\leq \frac{1}{2\sqrt{2}}|t|^3 + \frac{1}{2\sqrt{2}}|t|^3 = \frac{1}{\sqrt{2}}|t|^3,$$

as required.

Proposition 5.4 *The only integral solutions $x, y \in \mathbf{Z}$ of the equation*

$$y^2 = x^4 + x^3 + 1$$

are given by $(x, y) = (-2, \pm 3), (-1, \pm 1), (0, \pm 1),$ and $(2, \pm 5)$.

Proof Consider the algebraic curve given by the equation $Y^2 = X^4 + X^3 + 1$. The associated homogeneous equation $Y^2 Z^2 = X^4 + X^3 Z + Z^4$ describes a projective irreducible curve in the projective plane $\mathbf{P}^2$. This curve has a singularity at infinity. An equation that describes a curve with the same function field, but that is smooth in a neighborhood of infinity, is given by $T^2 = 1 + S + S^4$, where $T = Y/X^2$ and $S = 1/X$. The points at infinity are the points for which $S = 0$. We let Q denote the point corresponding to $T = 1$ and Q' the point with $T = -1$. A point (x, y) on the curve approaches Q when $y \to +\infty$, while it approaches Q' when $y \to -\infty$.

We consider the function Y. By Gauss's lemma, a rational point $P = (x, y)$ on the curve is Y-integral precisely when it is integral in the usual sense: It means $x, y \in \mathbf{Z}$. The points Q and Q' are the only poles of the function Y. As a consequence, the points Q and Q' are also the only poles of the function X.

The function

$$Y - X^2\left(1 + \frac{1}{2X} - \frac{1}{8X^2}\right) = Y - X^2 - \frac{1}{2}X + \frac{1}{8}$$

is a polynomial in X and Y. Therefore, its poles can only be Q or Q'. The function remains bounded as the point (x, y) approaches the pole Q or, equivalently, when $|x|$ and $y \to +\infty$. Indeed, since

$$y^2 = x^4 + x^3 + 1 = x^4\left(1 + \frac{1}{x} + \frac{1}{x^4}\right),$$

Lemma 5.3 implies that

$$\left|y - x^2 - \frac{1}{2}x + \frac{1}{8}\right| = |x|^2\left|F\left(\frac{1}{x}\right) - F_2\left(\frac{1}{x}\right)\right| \le \frac{1}{\sqrt{2}|x|}$$

whenever $|x| \ge 2$. Therefore, the only pole is Q'. Multiplying by 8, we obtain the function

$$Z = 8Y - 8X^2 - 4X + 1.$$

It is integral over $\mathbf{Z}[X]$ and hence has integral values in integral points of the curve $Y^2 = X^4 + X^3 + 1$. Since the function Z attains *odd* values in integral points, it does not vanish. It follows that we have

$$1 \le |Z(P)| = \left|8y - 8x^2 - 4x + 1\right| \le \frac{4\sqrt{2}}{|x|}.$$

This implies that when $x, y \in \mathbf{Z}$ satisfy $y^2 = x^4 + x^3 + 1$, we must have $|x| < 4\sqrt{2} = 5.6\ldots$. Checking all possible values $-5 \le x \le 5$, we find that $x^4 + x^3 + 1$ is a square precisely when $x = -2, -1, 0, 2$. This proves the proposition.

Non-example 5.5 (The Pell equation) Let d be a natural number. The curve given by the equation $X^2 - dY^2 = 1$ has two points, Q and Q', at infinity. They become visible when we consider the associated projective curve C in $\mathbf{P}^2$ given by the homogeneous equation $X^2 - dY^2 = Z^2$. Putting $T = X/Y$, they are given by $T = \pm\sqrt{d}$. The points Q and Q' are the poles of the functions X and Y. If d is a square, Q and Q' are *not* conjugate over $\mathbf{Q}$ and Runge's method applies. The sets $U = \{(x, y) \in C(\mathbf{R}) : xy > 0\}$ and $U' = \{(x, y) \in C(\mathbf{R}) : xy < 0\}$ are disjoint open neighborhoods of Q and Q', respectively.

Actually, it is easy to see directly that when D is a square, the only integral solutions to the equation $x^2 - dy^2 = 1$ are given by $x = \pm 1$ and $y = 0$. On the other hand, if D is *not* a square, we obtain a *Pell equation*. In this case, Runge's method does *not* imply that there are only finitely many solutions $x, y \in \mathbf{Z}$. This is explained by the fact that the two poles are *conjugate* over $\mathbf{Q}$. This is a good thing, because it is well known that the Pell equation $X^2 - dY^2 = 1$ admits *infinitely many* solutions $x, y \in \mathbf{Z}$.

In various parts of the proof of Catalan's conjecture, Runge's method is applied to *families* of curves. The curves and the bounds one obtains typically depend on the exponents p and q of Catalan's equation. It is essential that one controls this dependence. We illustrate the method in a simple case.

We make use of the following lemma. Let $q \geq 2$ be an integer and let $F(t)$ be the function of a real variable $t > 0$ given by

$$F(t) = \left(\frac{1+t^q}{1+t}\right)^{1/2}.$$

Lemma 5.6 *Let $F_m(t)$ denote the sum of the terms of degree at most m of the Taylor series around 0 of $F(t)$. For $m < q$, we have*

$$|F(t) - F_m(t)| < t^{m+1} \text{ for } t \in \mathbf{R} \text{ with } 0 < t < 1.$$

Proof Since we have $m < q$, the polynomial $F_m(t)$ is also the sum of the terms of degree at most m of the Taylor series expansion around 0 of the function $(1+t)^{-1/2}$. It is convenient to use the latter function when doing the estimates. When $0 < t < 1$, we have

$$|F(t) - F_m(t)| \leq |F(t) - (1+t)^{-1/2}| + |(1+t)^{-1/2} - F_m(t)|$$

$$\leq \frac{\frac{1}{2}t^q}{(1+t)^{1/2}} + \left|\binom{-1/2}{m}\right| \frac{t^{m+1}}{(1+t)^{m+3/2}}.$$

The estimate of the first term is elementary. The second term is estimated by applying Lemma 5.2 to the function $(1+t)^{-1/2}$. By Exercise 5.6, the absolute value of the binomial coefficient is at most $1/2$. This leads to

$$|F(t) - F_m(t)| \leq t^{m+1}\left(\frac{1}{2}t^{q-1-m} + \frac{1}{2}\right) < t^{m+1},$$

as required.

We give an alternative proof of Lemma 3.2. In other words, we show the following.

Proposition 5.7 *Let $q \geq 3$ be prime and suppose x, y are nonzero integers satisfying the equation $x^2 - y^q = 1$. Then we have $x \equiv 0 \pmod{q}$.*

Proof Let $x, y \in \mathbf{Z}$ be a nonzero solution and suppose $x \not\equiv 0 \pmod{q}$. We have

$$(y+1)\left(\frac{y^q+1}{y+1}\right) = x^2.$$

By Exercise 3.2, the gcd of the two factors on the left divides q. Since $x \not\equiv 0 \pmod{q}$, the gcd is therefore equal to 1. By Exercise 2.3, each factor is then itself a square times ± 1. Since $y^q = x^2 - 1$, the number Y is positive. This implies that both factors $y + 1$ and $(y^q + 1)/(y + 1)$ are positive, so that we actually have plus signs. Therefore, we have for certain integers $u, v \in \mathbf{Z}$,

$$y + 1 = u^2,$$
$$\frac{y^q + 1}{y + 1} = v^2.$$

The point (y, v) lies on the algebraic curve given by the equation

$$Y^{q-1} - Y^{q-2} + \ldots - Y + 1 = V^2.$$

The algebraic curve described by this equation is irreducible. The associated homogeneous equation $Y^{q-1} - Y^{q-2}Z + \ldots - YZ^{q-1} + Z^{q-1} = V^2 Z^{q-3}$ describes a projective irreducible curve in $\mathbf{P}^2$. This curve has a singularity at infinity. An equation that describes a curve with the same function field, but that is smooth in a neighborhood of infinity, is given by $W^2 = 1 - U + \ldots + U^{q-1}$, where $W = V/Y^{(q-1)/2}, U = 1/Y$. This shows that the curve C has two points, Q and Q', at infinity corresponding to $W = -1$ and $W = 1$, respectively. The function on C with respect to which we define the integrality is Y. In other words, a rational point (y, v) is integral when $y \in \mathbf{Z}$. Gauss's lemma implies that then v is also in $\mathbf{Z}$. The function Y has two zeroes. Therefore, it has two poles—the points Q and Q'.

We want to show that there are no rational points (y, v) on the curve C with $y > 0$. By symmetry, we may assume that $v > 0$ as well. The subset $U = \{(y, v) \in C(\mathbf{R}) \text{ with } vy^{(q-1)/2} < 0\} \cup \{Q\}$ is an open neighborhood in $C(\mathbf{R})$ of the pole Q. All rational points (y, v) of C with $y, v > 0$ are contained in the complement of U. Proposition 5.1 implies that there are at most finitely many integral points (y,z) outside the open set U.

We now apply Runge's method. We exhibit a function Z without any poles outside Q. Set $m = (q - 1)/2$ and consider the function

$$Z = V - Y^m F_m\left(\frac{1}{Y}\right),$$

where $F_m(t)$ is the polynomial that occurs in Lemma 5.6. We have

$$Z = V - Y^m + \frac{1}{2}Y^{m-1} + \ldots - \binom{-\frac{1}{2}}{m-1}Y - \binom{-\frac{1}{2}}{m}.$$

Since neither function Y nor V has any poles outside Q and Q', the same is true for Z. Next we show that $Z(P)$ tends to zero when P approaches Q'. Let $P = (y, v)$ with $y, v > 0$. Then we have

$$v = \left(\frac{y^q + 1}{y + 1} \right)^{1/2} = y^m F \left(\frac{1}{y} \right).$$

By Lemma 3.1 *(ii)*, we have $y \geq 2^{q-1} - 2 \geq 2$. Therefore, Lemma 5.6 implies

$$|Z(P)| = \left| v - y^m F_m \left(\frac{1}{y} \right) \right| = \left| y^m \left(F \left(\frac{1}{y} \right) - F_m \left(\frac{1}{y} \right) \right) \right|$$

$$< \frac{|y|^m}{y^{m+1}} = \frac{1}{y}.$$

It follows that $Z(P)$ tends to zero when $y \to +\infty$, i.e., when $P = (y, v)$ approaches Q'. Therefore, the function Z has no pole at Q'.

The function Z introduced above is integral over $\mathbf{Q}[Y]$, but not over the ring $\mathbf{Z}[Y]$. This follows from the fact that the coefficients of $F_m(t)$ are not in $\mathbf{Z}$. Indeed, by Exercise 5.7, the binomial coefficients $\binom{-\frac{1}{2}}{k}$ have denominators equal to $2^{k + \mathrm{ord}_2(k!)}$. Therefore, we set $D = 2^{m + \mathrm{ord}_2(m!)}$ and consider the function $Z' = DZ$. This is a polynomial in Y and V with coefficients in $\mathbf{Z}$. This is the function that occurs in the description of Runge's method given above. It is integral over $\mathbf{Z}[Y]$ and therefore takes integral values in all integral points P of the curve C. We have the following estimate for $Z'(P) = DZ(P)$ when $P = (y, v)$ is an integral point with $y, v > 0$:

$$|Z'(P)| = D|Z(P)| < \frac{D}{y}.$$

We use additional information to complete the proof of the proposition. By Exercise 5.5, we have $\mathrm{ord}_2(m!) < m$ and hence $D \leq 2^{m+(m-1)} = 2^{q-2}$. By Lemma 3.1 *(ii)*, we have $y \geq 2^{q-1} - 2$. This implies

$$|Z'(P)| < \frac{2^{q-2}}{2^{q-1} - 2},$$

which is at most 1 because we have $q \geq 3$. Since $Z'(P)$ is an integer, it follows that $Z'(P) = 0$. This implies

$$Dv = \sum_{k=1}^{m} D \binom{-\frac{1}{2}}{k} y^{m-k}.$$

Since D is even, so is the number on the left. On the other hand, Exercise 5.7 implies that for $0 < k < m$, the number $D\binom{-\frac{1}{2}}{k}$ is an *even* integer, while for $k = m$, it is *odd*. This is impossible.

This contradiction shows that there are no integral points $P = (y, v)$ with $y \neq 0$. Therefore, our assumption is wrong and we must have $x \equiv 0 \pmod{q}$, as required.

Runge's method extends naturally to curves over algebraic number fields. Let F be a number field and let O_F denote its ring of integers. Let C be a smooth, projective, absolutely irreducible algebraic curve over F and let Y be a function on C that is defined over F. An F-rational point $P \in C(F)$ is called *integral* if the value of the function Y in the point P is in O_F. Let Q be a pole of Y and let Z be a function in $F(C)$ that has no poles outside the set $Q = Q_1, \ldots, Q_r$ of $\mathrm{Gal}(\overline{F}/F)$-conjugates of Q. By multiplying Z by a suitable nonzero integer, we may assume that it is integral over $O_F[Y]$. For each embedding $\phi : F \subset \mathbf{C}$, one can effectively bound the values of the function $|\phi(Z(P))|$ in the points P that are outside a given open neighborhood U of $\phi(Q_1), \ldots, \phi(Q_r)$ in the Riemann surface $C(\mathbf{C})$.

In certain situations, one can use this approach and show that the set of *all* integral points $P \in C(F)$ is finite. Suppose, for instance, that the poles of the function Y are, up to $\mathrm{Gal}(\overline{F}/F)$-conjugacy, given by $Q^{(1)}, \ldots, Q^{(r)}$. For each $i = 1, \ldots, r$, let Z_i be a function in $F(C)$ that is integral over the ring $F[Y]$ and has no poles outside $Q^{(i)}$. Then it also has no poles in any of the $\mathrm{Gal}(\overline{F}/F)$-conjugates of $Q^{(i)}$. For every embedding $\phi : F \hookrightarrow \mathbf{C}$, and each $i = 1, \ldots, r$, let $U_i \subset C(\mathbf{C})$ be a neighborhood of the point $\phi(Q^{(i)})$ and its $\mathrm{Gal}(\overline{F}/F)$-conjugates. In addition, suppose that the U_i are disjoint. Then for any point P in $C(F)$, the point $\phi(P) \in C(\mathbf{C})$ is contained in at most one set U_i. It follows that the absolute values of the complex numbers $\phi(Z_j(P)) \in \mathbf{C}$ can be bounded for each $j \neq i$.

Therefore, if the number r of poles of Y *exceeds* the number of complex embeddings of F into $\mathbf{C}$ (taken up to complex conjugation), the box principle implies that for each point P, at least one function Z_j has the property that *all* complex numbers $\phi(Z_j(P))$ are bounded. When $Z_j(P)$ is in O_F, this gives us an explicit finite list of possible values $Z_j(P)$. From this it is easy to determine all integral points $P \in C(F)$.

Another favorable situation occurs when F is totally real and Y has a pole Q whose conjugates are all complex rather than real. In this case all points $P \in C(F)$ are far away from Q and its Galois conjugates with respect to every embedding $\phi : F \hookrightarrow \mathbf{C}$. The latter situation occurs in Mihăilescu's proof of Theorem 12.4.

Exercises

5.1 Let $R \subset S$ be an extension of rings. Let $x \in S$. Show that the following are equivalent:

    (a) the element x is integral over R;
    (b) we have $xM \subset M$ for some finitely generated R-submodule M of the additive group of S;
    (c) the subring $R[x]$ of S is finite over R.

5.2 Let R be an integrally closed domain with quotient field K and let L be a finite extension of K. Let $x \in L$. Then the following are equivalent:

    (a) the element x is integral over R;
    (b) the minimum polynomial of x over K is contained in $R[X]$;

(c) the characteristic polynomial of x is contained in $R[X]$. (The characteristic polynomial of x is the characteristic polynomial of the matrix of the K-linear map $L \longrightarrow L$ that is given by multiplication by x.)

5.3 Let F be a field and let C be a smooth, irreducible, projective curve over F. Let Y be a nonconstant element of the function field $F(C)$ and let D be its pole divisor. Let g be a function in $F(C)$ that has no poles outside D. Show that g is integral over $F[Y]$. (Hint: See [1, Cor. 5.22].)

5.4 Let R be a Noetherian integrally closed domain with quotient field K of characteristic zero. Let L be a field extension of K of finite degree. Show that the integral closure of R in L is finite over R. (Hint: See [1, Prop. 5.17].)

5.5 Let $n \geq 1$ be an integer and let P be a prime. For any $x \in \mathbf{R}$, we let $[x]$ denote the *integral part* of x.

(a) Show that $\mathrm{ord}_p(n!) = \sum_{i=1}^{\infty} [\frac{n}{p^i}]$.

(b) Show that $\mathrm{ord}_p(n!) = \frac{n-s(n)}{p-1}$, where $s(n)$ is the number of digits of n written in base p. Conclude that $\mathrm{ord}_p(n!) < \frac{n}{p-1}$.

5.6 For any domain R of characteristic zero and any $x \in R$ and $k \in \mathbf{Z}_{\geq 0}$, we define the binomial coefficient $\binom{x}{k}$ in the quotient field of R as follows. We put $\binom{x}{k} = 1$ when $k = 0$ and we let

$$\binom{x}{k} = \frac{x(x-1) \cdot \ldots \cdot (x-(k-1))}{1 \cdot 2 \cdot \ldots \cdot k}, \quad \text{for } k \geq 1.$$

Let $k \in \mathbf{Z}_{>0}$. Show that for $x \in \mathbf{R}$, one has

(a)

$$\left| \binom{x}{k} \right| \leq \frac{1}{k} \quad \text{for } 0 \leq x \leq k-1;$$

(b)

$$\left| \binom{x}{k} \right| \leq |x| \quad \text{for } -1 \leq x \leq 0.$$

5.7 (a) Let l be a prime and let x be contained in the ring $\mathbf{Z}_l$ of l-adic integers. Show that $\binom{x}{k}$ is contained in $\mathbf{Z}_l$ for every $k \in \mathbf{Z}_{\geq 0}$. (Hint: The function $x \mapsto \binom{x}{k}$ is a continuous map from the field of l-adic numbers $\mathbf{Q}_l$ to itself. Now use Exercise 2.4 and the fact that $\mathbf{Z}$ is dense in $\mathbf{Z}_l$.)

(b) Let $x \in \mathbf{Q}^*$. Show that for every natural number k, the binomial coefficient $\binom{x}{k}$ is contained in the ring $\mathbf{Z}[x]$. In other words, show that $\binom{x}{k}$ is l-integral for all primes l that do not divide the denominator of x. [Hint: Use part (a).]

(c) Let $x \in \mathbf{Q}^*$ and suppose that the denominator of x is a prime number q. Show that the denominator of $\binom{x}{k}$ is equal to $q^{k + \mathrm{ord}_q(k!)}$.

5.8 Let $\alpha \in \mathbf{R}$ and $F(t) = (1+t)^\alpha$. Let $m > \alpha$ and let $t \in \mathbf{C}$ with $|t| < 1$. Using the notation of Lemma 5.2, show that

$$|F(t) - F_m(t)| \leq \left|\binom{\alpha}{m+1}\right| \frac{|t|^{m+1}}{(1-|t|)^{m+1-\alpha}}.$$

5.9 Let $F, G \in \mathbf{Z}[X]$ be two monic polynomials of degree $n \geq 2$. Let C be the projective curve given by the homogeneous equation $Z^n F(X/Z) = Z^n G(Y/Z)$ in the projective plane $\mathbf{P}^2$.

(a) Show that the infinite points of the algebraic curve C are the points characterized by the equation $T^n = 1$, where $T = Y/X$.

(b) Show that the function $X - Y$ in the function field $\mathbf{Q}(C)$ has no pole at the infinite point given by $T = 1$.

(c) Bound the size of the integral points on C in terms of the polynomials F and G.

(d) Find all integral solutions of the equation $X^3 - X + 1 = Y^3 + 2Y^2$.

6
Cassels' theorem

In this chapter, we prove J.W.S. Cassels' theorem (Fig. 6.1). See [8]. For the sake of reference, we prove the following easy lemma.

Lemma 6.1 *Let $p, q \geq 2$ be integers and suppose that x,y are nonzero integers that are a solution to the equation $x^p - y^q = 1$. Then p and q are necessarily distinct.*

Proof Indeed, if p were equal to q, then we would have $x^p - y^p = 1$. But pth powers are far apart. When the distance between them is 1, necessarily one of x,y is zero. See also Exercise 3.3. Therefore, we have $p \neq q$, as required.

When p and q are both odd, there is a symmetry: If (x, y, p, q) is a solution to Catalan's equation, then so is $(-y, -x, q, p)$. Therefore, we may assume $p > q$.

Cassels' theorem is concerned with Catalan's equation for the odd prime exponents p and q. We first prove the easy part of his result.

Proposition 6.2 *Let $p > q$ be two odd primes and suppose that x,y are nonzero integers for which $x^p - y^q = 1$. Then both of the following hold:*

(i) q divides x;
(ii) $|x| \geq q + q^{p-1}$.

Proof (i) Suppose that q does not divide x. By Exercise 3.2, the two factors $y + 1$ and $(y^q + 1)/(y + 1)$ of $y^q + 1 = x^p$ are coprime. Since their product is a pth power, Exercise 2.3 implies that we have

$$y + 1 = b^p,$$
$$\frac{y^q + 1}{y + 1} = u^p$$

for certain $b, u \in \mathbf{Z}$. It follows that $x^p - (b^p - 1)^q = 1$. This implies that the two pth powers x^p and b^{pq} are relatively close to one another. Too close, as we show now. As a function of x, the expression $x^p - (b^p - 1)^q$ is strictly increasing. When $b > 0$, substituting $x = b^q$ gives a value exceeding 1, while substituting $x = b^q - 1$ leads to a negative value. Indeed, we have $(b^q - 1)^p < (b^p - 1)^q$ because $p > q$. On the other hand, when $b < 0$, substituting $x = b^q - 1$ gives a

R. Schoof, *Catalan's Conjecture*, DOI: 10.1007/978-1-84800-185-5_6,
© Springer-Verlag London Limited 2008

Fig. 6.1 J.W.S. Cassels. (Reproduced from *I Have a Photographic Memory* by Paul R. Halmos. ©
American Mathematical Society, 1993, with permission from the American Mathematical Society.)

value exceeding 1, while substituting $x = b^q$ leads to a negative value. Indeed, we
have $(-b^p + 1)^q < (-b^q + 1)^p$ because $p > q$. See Exercise 6.1 for the inequalities.
In either case, an *integer* x for which the expression $x^p - (b^p - 1)^q$ is equal to 1
does not exist.

We conclude that q divides x, as required.

(ii) Part *(i)* implies that the gcd of the factors $y+1$ and $(y^q + 1)/(y + 1)$ is equal
to q. Exercise 6.2 implies that $(y^q + 1)/(y + 1) \equiv q \pmod{q^2}$. It follows that q^2
does not divide $(y^q + 1)/(y + 1)$. Since q^p divides $y^q + 1$, it must be the case that
q^{p-1} divides $y + 1$. It follows that the numbers

$$\frac{y^q + 1}{q(y + 1)} \quad \text{and} \quad \frac{y + 1}{q^{p-1}}$$

are coprime integers. Since their product is a pth power, Exercise 2.3 tells us then
that

$$y + 1 = q^{p-1}b^p,$$
$$\frac{y^q + 1}{y + 1} = qu^p,$$
$$x = qub.$$

for certain $b, u \in \mathbf{Z}$. Note that u is positive by Exercise 3.1. Since Exercise 6.2 implies

$$\frac{y^q + 1}{y + 1} \equiv q \ (\mathrm{mod}(y + 1)),$$

we have $qu^p \equiv q \ (\mathrm{mod}\ q^{p-1})$ and hence $u^p \equiv 1 \ (\mathrm{mod}\ q^{p-2})$. Since p exceeds q, it does not divide the order of the group $(\mathbf{Z}/q^{p-2}\mathbf{Z})^*$. It follows that $u \equiv 1 \ (\mathrm{mod}\ q^{p-2})$. We claim that $u \neq 1$. Indeed, if $u = 1$, we have $q = qu^p = (y^q + 1)/(y + 1)$. Since $x \neq 0$ and hence $y \neq -1$, Exercise 6.4 implies that $q = 3$ and $y = 2$. This means that $x^p = y^q + 1 = 9$, which is impossible. From the fact that $u > 0$, it follows that $u \geq 1 + q^{p-2}$. Since we have $x = qub$, part *(ii)* follows.

The proof of Proposition 6.2 *(i)* is a very easy instance of Runge's method. The relevant curve is given by $X^p - (B^p - 1)^q = 1$. The points at infinity are given by $T^p = 1$, where T is the function X/B^q. When $p > q$, the function $Z = B^q - X$ is zero in the infinite point Q given by $T = 1$. Therefore, Z has no poles outside the infinite point Q' given by $T^{p-1} + \ldots + T + 1 = 0$. The estimates that appear in the proof show precisely that the values of the function Z on the real points of the curve are in the half-open interval $[0, 1)$. This implies that Z vanishes on integral points. It easily follows that the only integral points on the curve are given by $(X, B) = (0, 0)$ and $(1, 1)$.

The following lemma is used in the proof of Cassels' theorem. For odd integers $p > q$, let $F(t)$ denote the function of a real variable t given by

$$F(t) = \left((1 + t)^p - t^p\right)^{1/q}.$$

Recall that $[x]$ denotes the integral part of $x \in \mathbf{R}$.

Lemma 6.3 *Let p and q be odd integers satisfying $p > q \geq 3$, let $m = [\frac{p}{q}] + 1$, and let $F_m(t)$ denote the sum of the terms of degree at most m of the Taylor series around 0 of the function $F(t)$. Then we have*

$$|F(t) - F_m(t)| \quad \leq \quad \frac{|t|^{m+1}}{(1 - |t|)^2}, \quad \text{for } t \in \mathbf{R} \text{ with } |t| < 1.$$

Proof Since we have $m < p$, the polynomial $F_m(t)$ is also the sum of the terms of degree $\leq m$ of the Taylor series of the function $(1 + t)^{p/q}$ around 0. It is convenient to do the estimates using the latter function. For $t \in \mathbf{R}$ satisfying $|t| < 1$, we have

$$|F(t) - F_m(t)| \quad \leq \quad |F(t) - (1 + t)^{p/q}| + |(1 + t)^{p/q} - F_m(t)|.$$

We estimate the first term as follows:

$$|F(t) - (1 + t)^{p/q}| \quad \leq \quad \frac{1}{q}|t|^p|t'|^{\frac{1}{q}-1}$$

$$\leq \quad \frac{1}{q}|t|^p(1 - |t|)^{p(\frac{1}{q}-1)}$$

$$\leq \quad \frac{1}{q}\frac{|t|^p}{q(1 - |t|)^2}.$$

The first inequality is an application of the mean value theorem to the function $x \mapsto x^{1/q}$. Here t' is a real number between $(1+t)^p$ and $(1+t)^p - t^p$. Exercise 6.5 shows that $|t'| \geq (1-|t|)^p$, which implies the second inequality. The third inequality follows from the fact that $p(1 - \frac{1}{q}) > 2$.

We estimate the second term as follows:

$$|(1+t)^{p/q} - F_m(t)| \leq |t|^{m+1} \left| \binom{\frac{p}{q}}{m+1} \right| (1 - |t|)^{-m-1+p/q}$$

$$\leq \frac{1}{m+1} \frac{|t|^{m+1}}{(1 - |t|)^2}.$$

The first inequality follows from Lemma 5.2 or Exercise 5.8. By Exercise 5.6(a), the binomial coefficient is at most $1/(m+1)$. The second inequality follows from this and the fact that $\frac{p}{q} - m - 1 \geq -2$.

Combining the inequalities, we find

$$|F(t) - F_m(t)| \leq \left(\frac{|t|^p}{q} + \frac{|t|^{m+1}}{m+1} \right) \frac{1}{(1 - |t|)^2}.$$

Since $p > m + 1$ and both m and q are at least 2, the right-hand side is at most $|t|^{m+1}/(1 - |t|)^2$, as required. This proves the lemma.

Theorem 6.4 *(J.W.S. Cassels, 1960) Let p,q be two odd primes, and suppose that x,y are two nonzero integers for which $x^p - y^q = 1$. Then q divides x and p divides y.*

Proof By Lemma 6.1, we have $p \neq q$. By symmetry, we may assume $p > q$. The fact that the "small" prime q divides x was proved in Proposition 6.2. Now we show that the "large" prime p divides y. The proof is an application of Runge's method.

The proof is by contradiction. Suppose that p does not divide y. By Exercise 3.2, the numbers $x - 1$ and $(x^p - 1)/(x - 1)$ are then coprime and Exercise 2.3 therefore implies

$$x - 1 = a^q,$$

for a certain nonzero $a \in \mathbf{Z}$. Substituting this into Catalan's equation, we obtain

$$y^q = (a^q + 1)^p - 1.$$

Writing $F(t)$ for the function $((1+t)^p - t^p)^{1/q}$ (for $t \in \mathbf{R}$), this means that we have $y = a^p F(\frac{1}{a^q})$. As in Lemma 6.3, we set $m = [\frac{p}{q}] + 1$. We set

$$z = a^{mq-p}y - a^{mq} F_m \left(\frac{1}{a^q} \right).$$

Since the exponent $mq - p$ is positive, the number z is equal to a polynomial expression in a and y. The coefficients of the polynomial

$$t^m F_m \left(\frac{1}{t} \right) = t^m + \binom{\frac{p}{q}}{1} t^{m-1} + \ldots + \binom{\frac{p}{q}}{m}$$

are not integral. Indeed, they are the binomial coefficients $\binom{\frac{p}{q}}{k}$ for $0 \le k \le m$. By Exercise 5.7, the denominators of these numbers are equal to $q^{k + \mathrm{ord}_q(k!)}$. Therefore, each denominator divides $D = q^{m + \mathrm{ord}_q(m!)}$, and the number

$$Dz = Da^{mq-p}y - a^{mq} \sum_{k=0}^{m} D \binom{\frac{p}{q}}{k} a^{-qk}$$

is an integer. In addition, Dz *is not zero*, because it is not congruent to zero modulo q. This follows from the fact that all terms in the sum except the mth one are integers that are divisible by q. Indeed, the mth coefficient is equal to the binomial coefficient $\binom{\frac{p}{q}}{m}$ multiplied by its denominator $D = q^{m + \mathrm{ord}_q(m!)}$.

We conclude the proof by estimating the size of Dz. We have

$$z = a^{mq} \left(F \left(\frac{1}{a^q} \right) - F_m \left(\frac{1}{a^q} \right) \right).$$

Since $x \ne 0$ but is congruent to 0 (mod q), we have $a \ne 0, \pm 1$ and hence $|a| \ge 2$. Therefore, Lemma 6.3 with $t = 1/a^q$ applies. We have

$$|z| \quad \le \quad \frac{|a|^q}{(|a|^q - 1)^2} \quad \le \quad \frac{1}{|a|^q - 2} \quad \le \quad \frac{1}{|x| - 3}.$$

By Proposition 6.2 *(ii)*, we have $|x| \ge q^{p-1} + q$ and hence $|x| - 3 \ge q^{p-1}$. Therefore,

$$|Dz| \quad \le \quad \frac{D}{|x| - 3} \quad \le \quad q^{m + \mathrm{ord}_q(m!) - (p-1)}.$$

We now come to the critical part of the estimate. It is essential that we show that the exponent is negative. By Exercise 5.5(b), we have

$$m + \mathrm{ord}_q(m!) - (p - 1) \le m \left(1 + \frac{1}{q-1} \right) - (p - 1).$$

Since $m < \frac{p}{q} + 1$, the right-hand side is *strictly* smaller than

$$\left(\frac{p}{q} + 1 \right) \left(1 + \frac{1}{q-1} \right) - (p - 1) \quad = \quad \frac{3 - (p-2)(q-2)}{q - 1}.$$

and is ≤ 0 because we have $q \ge 3$ and $p \ge 5$.

It follows that the absolute value of the nonzero integer Dz is strictly smaller than 1. This contradiction proves the theorem.

In terms of the description of Runge's method given in chapter 5, the algebraic curve C that is relevant for the proof of Cassels' theorem is given by the affine equation

$$Y^q = (1 + A^q)^p - 1 = A^{qp} + pA^{q(p-1)} + \ldots + pA^q.$$

By Gauss's lemma, the integral points $P = (a, y) \in C(\mathbf{Q})$ are precisely the rational points at which the function A assumes integral values. The pole divisor of A is the divisor of infinite points of C. Setting $W = Y/A^p$, we have

$$W^q = 1 + \frac{p}{A^q} + \ldots + \frac{p}{A^{(p-1)q}},$$

and we see that the pole divisor of A is given by the equation $W^q = 1$. Therefore, A admits, up to Galois conjugacy, two poles: The pole Q' is given by $W = 1$ and is defined over $\mathbf{Q}$, while the pole Q consists of the Galois conjugates of the point given by $W = \zeta_q$, where ζ_q denotes a primitive Qth root of unity. The pole q is defined over the cyclotomic field $\mathbf{Q}(\zeta_q)$. Since Q is not defined over $\mathbf{R}$, the rational points on the curve C are all "far away" from it. The discussion above implies that the function

$$Z = A^{mq-p} y - A^{mq} F_m \left(\frac{1}{A^q} \right)$$

admits no poles outside Q. Indeed, since $mq - p > 0$, the function Z is a polynomial in the functions A and Y. Since both A and Y do not have any poles outside Q and Q', neither does Z. The estimate shows that $Z(P)$ tends to zero when P tends to Q'. Therefore, Q' is not a pole of Z either. The function Z is integral over $\mathbf{Q}[A]$ but not over $\mathbf{Z}[A]$. This is caused by the denominators of the coefficients of $F_m(t)$. The function $Z' = DZ$ is integral over the polynomial ring $\mathbf{Z}[A]$. Therefore, it takes integral values in the integral points in $C(\mathbf{Q})$. As we have seen in the proof of Theorem 6.4, this leads to the determination of all integral points.

An essential point of Cassels' proof is that in the application of Runge's method, one is able to do the estimates *uniformly* in the exponents p and q.

Corollary 6.5 *Let p, q be odd primes and let x, y be nonzero integers satisfying Catalan's equation $x^p - y^q = 1$. Then*

(i) there exist $a, v \in \mathbf{Z}$ such that

$$y = pav, \quad x - 1 = p^{q-1} a^q, \quad \frac{x^p - 1}{x - 1} = pv^q;$$

(ii) there exist $b, u \in \mathbf{Z}$ such that

$$x = qbu, \quad y + 1 = q^{p-1} b^p, \quad \frac{y^q + 1}{y + 1} = qu^p;$$

(iii) we have

$$|x| \geq \max(p^{q-1} - 1, q^{p-1} + q), \quad |y| \geq \max(q^{p-1} - 1, p^{q-1} + p).$$

Proof Since p divides y, we have $x^p \equiv 1 \pmod{p}$ and hence $x \equiv 1 \pmod{p}$. Since p^q divides $x^p - 1$ and since $(x^p - 1)/(x - 1) \equiv p \pmod{p^2}$ by Exercise 6.2, we see that p^{q-1} divides $x - 1$. An application of Exercise 2.3 to the factorization

$$\frac{x - 1}{p^{q-1}} \cdot \frac{x^p - 1}{p(x - 1)} = \left(\frac{y}{p}\right)^q$$

implies *(i)*. Part *(ii)* follows by symmetry. It suffices to replace the quadruple (x, y, p, q) by $(-y, -x, q, p)$. Note that both u and v are positive. For part *(iii)*, note that $a \neq 0$, so that $|x| \geq p^{q-1} - 1$ by part *(i)*. If $p < q$, Exercise 6.3 implies that this bound exceeds $q^{p-1} + q$ and we are done. In the other case, Proposition 6.2 *(ii)* implies $|x| \geq q^{p-1} + q$, as required. The inequalities for y follow by the symmetry $(x, y, p, q) \leftrightarrow (-y, -x, q, p)$.

For a nonzero solution $x, y \in \mathbf{Z}$ to Catalan's equation, Corollary 6.5 provides us with three pieces of "local" information concerning a nonzero solution x,y to Catalan's equation $x^p - y^q = 1$. We have

- $x \equiv 1 \pmod{p^{q-1}}$ and $y \equiv 0 \pmod{p}$,   (p-adic),
- $x \equiv 0 \pmod{q}$ and $y \equiv -1 \pmod{q^{p-1}}$,   ($q$-adic),
- $|x|, |y| \geq \max(q^{p-1} - 1, p^{q-1} - 1)$,   (Archimedean).

The inequalities of part *(iii)* of Corollary 6.5. can rather easily be improved upon. This was done by S. Hyyrö in 1964. See [18] or Exercise 6.6.

Exercises

6.1 Let $b \in \mathbf{R}_{>1}$. Show that $t \mapsto (b^t + 1)^{1/t}$ is a decreasing function from $\mathbf{R}_{>0}$ to $\mathbf{R}_{>0}$. Show that $t \mapsto (b^t - 1)^{1/t}$ is an increasing function from $\mathbf{R}_{>0}$ to $\mathbf{R}_{>0}$.

6.2 Let q be an odd prime.

    (a) Let $a \in \mathbf{Z}$ be different from 1. Show that $(a^q - 1)/(a - 1) \equiv q \pmod{(a-1)}$.
    (b) Let $a \in \mathbf{Z}$ satisfy $a \equiv 1 \pmod{q}$. Show that $(a^q - 1)/(a - 1) \equiv q \pmod{q^2}$.

6.3 Let $p > q$ be odd primes. Show that $q^{p-1} - 1 > p^{q-1} + p$.

6.4 Let q be prime. Consider the equation

$$\frac{y^q + 1}{y + 1} = y^{q-1} - y^{q-2} + \ldots - y + 1 = q.$$

(a) For $q \geq 5$, show that the only solution to $y \in \mathbf{Z}$ is given by $y = -1$.

(b) Determine all solutions $y \in \mathbf{Z}$ for $q = 3$.

6.5 For any two real numbers $p, x > 1$, show that $(x + 1)^p - (x - 1)^p > 1$.

6.6 (S. Hyyrö [18]) Suppose that p, q are odd primes and that $x, y \in \mathbf{Z}$ satisfy
$x^p - y^q = 1$. Corollary 6.5 says that there exist $a, v \in \mathbf{Z}$ such that $y = pav$
and $x - 1 = p^{q-1}a^q$. Similarly, there exist $b, u \in \mathbf{Z}$ such that $x = qbu$ and
$y + 1 = q^{p-1}b^p$.

(a) Show that $a \equiv -1 \pmod{q}$ and $b \equiv 1 \pmod{p}$.

(b) Show that $a \neq -1$ and $b \neq 1$. Conclude that $|a| \geq q - 1$ and $|b| \geq p - 1$.

(c) Show

$$|x| \geq p^{q-1}(q - 1)^q - 1,$$
$$|y| \geq q^{p-1}(p - 1)^p - 1.$$

7
An Obstruction Group

Nonzero solutions to Catalan's equation $x^p - y^q = 1$ naturally give rise to elements in certain obstruction groups. Mihăilescu's theorems can be viewed as results about the size of the submodules generated by these elements. In this section, we introduce the notions and notations that are used in the remaining sections.

Let $p > 2$ be prime, let ζ_p denote a primitive root of unity, and let $\mathbf{Q}(\zeta_p)$ denote the pth cyclotomic field. The Galois group G of $\mathbf{Q}(\zeta_p)$ over $\mathbf{Q}$ is given by

$$G = \{\sigma_a : a \in (\mathbf{Z}/p\mathbf{Z})^*\},$$

where σ_a acts on ζ_p by raising it to the ath power. Note that σ_{-1} is the automorphism that, for every embedding $\phi : \mathbf{Q}(\zeta_p) \hookrightarrow \mathbf{C}$, is induced by complex conjugation. More precisely, we have $\phi(\sigma_{-1}(x)) = \overline{\phi(x)}$ for every embedding ϕ and every number $x \in \mathbf{Q}(\zeta_p)$. We usually write ι for the element $\sigma_{-1} \in G$. By Exercise 7.2, the ring $\mathbf{Z}[\zeta_p]$ is the ring of integers of $\mathbf{Q}(\zeta_p)$. Its ideal class group is denoted by Cl_p. The order of the finite group Cl_p is called the *class number* of $\mathbf{Q}(\zeta_p)$. It is denoted by h_p. We define the group E_p of p-units by

$$E_p = \mathbf{Z}[\zeta_p, \tfrac{1}{p}]^*.$$

By Exercise 7.2, the unique prime $\mathfrak{p}$ lying over p in the ring $\mathbf{Z}[\zeta_p]$ is generated by $1 - \zeta_p$. The group E_p is generated by the unit group $\mathbf{Z}[\zeta_p]^*$ and by the element $1 - \zeta_p$. All groups defined so far are modules over the group ring $\mathbf{Z}[G]$. By Exercise 7.3, there is an exact sequence of $\mathbf{Z}[G]$-modules

$$0 \longrightarrow \mathbf{Z}[\zeta_p]^* \longrightarrow E_p \overset{\mathrm{ord}_{\mathfrak{p}}}{\longrightarrow} \mathbf{Z} \longrightarrow 0,$$

where $\mathrm{ord}_{\mathfrak{p}}$ is the valuation associated to the prime ideal $\mathfrak{p}$. We sometimes use *exponential notation* for the action of $\mathbf{Z}[G]$ on multiplicative groups.

Let q be an odd prime not equal to p. Consider the $\mathbf{F}_q[G]$-module

$$H = \left\{\alpha \in \mathbf{Q}(\zeta_p)^* : \mathrm{ord}_{\mathfrak{r}}(\alpha) \equiv 0 \pmod{q} \text{ for all primes } \mathfrak{r} \neq \mathfrak{p}\right\}/\mathbf{Q}(\zeta_p)^{*q}.$$

R. Schoof, *Catalan's Conjecture*, DOI: 10.1007/978-1-84800-185-5_7,
© Springer-Verlag London Limited 2008

In other words, an element $\alpha \in \mathbf{Q}(\zeta_p)^*$ (modulo qth powers) is contained in H if and only if the principal ideal generated by α is equal to $\mathfrak{a}^q \mathfrak{p}^k$ for some $k \in \mathbf{Z}$ and some fractional ideal $\mathfrak{a}$. The $\mathbf{F}_q[G]$-module H is the obstruction group that was mentioned in chapter 1. It plays an important role in our presentation of Mihăilescu's proof.

There is a natural homomorphism ψ from H to the group $Cl_p[q]$ of ideal classes that are annihilated by q. It sends α to the class of the fractional ideal $\mathfrak{a}$ for which $(\alpha) = \mathfrak{a}^q \mathfrak{p}^k$ for some $k \in \mathbf{Z}$.

Lemma 7.1 *Let p, q be distinct odd prime numbers and let H be the $\mathbf{F}_q[G]$-module introduced above. Then we have the following.*

(i) *There is a natural exact sequence of $\mathbf{F}_q[G]$-modules*

$$0 \longrightarrow E_p/E_p^q \longrightarrow H \overset{\psi}{\longrightarrow} Cl_p[q] \longrightarrow 0.$$

(ii) *The automorphism ι acts trivially on the group E_p/E_p^q.*

Proof Since the prime $\mathfrak{p}$ over p is a principal $\mathbf{Z}[\zeta_p]$-ideal, the map ψ is well defined. It is easy to see that the natural map $E_p/E_p^q \longrightarrow H$ is injective and that its image is contained in the kernel of ψ. Any ideal $\mathfrak{a}$ with $\mathfrak{a}^q = (\alpha)$ is the image under ψ of $\alpha \in H$. This shows that ψ is surjective. Finally, if $(\alpha) = \mathfrak{p}^k$ for some $k \in \mathbf{Z}$, the number α is necessarily a p-unit. This shows that the sequence is exact. This proves *(i)*.

For any unit $u \in \mathbf{Z}[\zeta_p]^*$ and every embedding $\phi : \mathbf{Q}(\zeta_p) \hookrightarrow \mathbf{C}$, the complex number $\phi(\frac{u}{\iota(u)}) = \phi(u)/\overline{\phi(u)}$ has absolute value 1. Exercise 7.1 implies therefore that $u^{1-\iota} = \frac{u}{\iota(u)}$ is a root of unity. Similarly,

$$(1 - \zeta_p)^{1-\iota} = \frac{1 - \zeta_p}{1 - \iota(\zeta_p)} = \frac{1 - \zeta_p}{1 - \zeta_p^{-1}} = -\zeta_p$$

is a root of unity. The group E_p is generated by $\mathbf{Z}[\zeta_p]^*$ and by $1 - \zeta_p$. Therefore, any $\varepsilon \in E_p$ has the property that $\varepsilon^{1-\iota}$ is a root of unity. Since the roots of unity in $\mathbf{Q}(\zeta_p)$ have order dividing $2p$, they are qth powers. This shows that the class of ε in the quotient group E_p/E_p^q is ι-invariant. This proves *(ii)*.

The following proposition explains the relevance of the group H with respect to Catalan's conjecture.

Proposition 7.2 *Let p, q be distinct odd primes and suppose that x, y are nonzero integers that satisfy Catalan's equation $x^p - y^q = 1$. Then the class of the element $x - \zeta_p$ modulo $\mathbf{Q}(\zeta_p)^{*q}$ is contained in the obstruction group H.*

Proof We have

$$\prod_{\zeta \in \mu_p} (x - \zeta) = y^q.$$

Here μ_p denotes the group of pth roots of unity. Since the right-hand side is a qth power, so is the left-hand side. The gcd of any two of the factors $x - \zeta$ divides the prime ideal $\mathfrak{p} = (1 - \zeta_p)$ of $\mathbf{Z}(\zeta_p)$. Therefore, we have $\mathrm{ord}_{\mathfrak{r}}(x - \zeta_p) \equiv 0 \pmod q$ for all primes $\mathfrak{r} \neq \mathfrak{p}$ of $\mathbf{Z}[\zeta_p]$.

This proves the proposition.

Several of Mihăilescu's results translate readily into properties of the element $x - \zeta_p$ in the obstruction group H. This is also true for some of the results obtained earlier. For instance, an important consequence of Cassels' theorem can be translated into a statement regarding the group H.

Proposition 7.3 *Let $x, y \in \mathbf{Z}$ be a nonzero solution to Catalan's equation. Then the element $x - \zeta_p$ of H is not contained in the index-q submodule H_0 of H given by*

$$H_0 = \left\{\alpha \in \mathbf{Q}(\zeta_p)^* : \mathrm{ord}_{\mathfrak{r}}(\alpha) \equiv 0 \ (\mathrm{mod}\ q)\ \textit{for all primes}\ \mathfrak{r}\right\} / \mathbf{Q}(\zeta_p)^{*q}.$$

Proof Corollary 6.5 of Cassels' result implies $x \equiv 1 \pmod p$. Equivalently, it says that $x - \zeta_p$ is divisible by $\mathfrak{p} = (1 - \zeta_p)$. Since $\mathfrak{p}$ divides $x - \zeta_p$ exactly once, this implies that

$$\mathrm{ord}_{\mathfrak{p}}(x - \zeta_p) \not\equiv 0 \ (\mathrm{mod}\ q).$$

This proves the proposition.

Next we exploit the action of complex conjugation $\iota \in G$ on the obstruction group H.

For any $\mathbf{Z}[G]$-module M, we define its plus part M^+ by $M^+ = M^{1+\iota} = \{(1+\iota)m : m \in M\}$ and its minus part M^- by $M^- = M^{1-\iota} = \{(1 - \iota)m : m \in M\}$. Both M^+ and M^- are $\mathbf{Z}[G]$-modules. Writing G^+ for the quotient group $G/\langle\iota\rangle$, the plus part M^+ is even a module over $\mathbf{Z}[G^+]$. When q is an odd prime number and M is a $\mathbf{Z}[G]$-module, then $M[q] = \{m \in M : qm = 0\}$ is also a $\mathbf{Z}[G]$-module. We have $M[q]^+ = M^+[q]$ and $M[q]^- = M^-[q]$. Since q is odd, any $\mathbf{F}_q[G]$-module M is the direct product of M^- and M^+. In particular, the obstruction group H is an $\mathbf{F}_q[G]$-module and is a direct product of its plus part H^+ and its minus part H^-.

The fixed field of ι is $\mathbf{Q}(\zeta_p^+)$, where ζ_p^+ denotes $\zeta_p + \zeta_p^{-1} = \zeta_p + \iota(\zeta_p)$. The quotient group $G^+ = G/\langle\iota\rangle$ is the Galois group of $\mathbf{Q}(\zeta_p^+)$ over $\mathbf{Q}$. We denote the class group of the ring of integers of $\mathbf{Q}(\zeta_p^+)$ by Cl_p^+. The minus class group Cl_p^- is the quotient of the class group Cl_p of $\mathbf{Q}(\zeta_p)$ by the image of the natural homomorphism $Cl_p^+ \longrightarrow Cl_p$.

There is a slight ambiguity in this notation. By Exercise 7.5, the natural map $Cl_p^+ \longrightarrow Cl_p$ is injective and its image is *contained* in the plus part of Cl_p. The plus part may contain this image strictly. However, by Exercise 7.6, the index is at most a power of 2. Since q is odd, this implies that the natural homomorphism $Cl_p^+[q] \longrightarrow Cl_p[q]^+$ is an isomorphism. Similarly, Exercise 7.7 implies that the natural map $Cl_p[q]^- \longrightarrow Cl_p^-[q]$ is an isomorphism. Since we only consider the q-parts of Cl_p, Cl_p^+, and Cl_p^-, this abuse of notation should not lead to any confusion.

The order of Cl_p^+ is called the *plus class number* and is denoted by h_p^+. The *minus class number* h_p^- of $\mathbf{Q}(\zeta_p)$ is the order of Cl_p^-. By Exercise 7.5, the map $Cl_p^+ \longrightarrow Cl_p$ is injective. Therefore, we have $h_p = h_p^+ h_p^-$.

Proposition 7.4 *Let p, q be two distinct odd prime numbers.*

(i) The natural homomorphism

$$H^- \xrightarrow{\cong} Cl_p^-[q]$$

is an isomorphism;

(ii) taking plus parts of the modules in the exact sequence of Lemma 7.1, we obtain the exact sequence

$$0 \longrightarrow E_p/E_p^q \longrightarrow H^+ \longrightarrow Cl_p^+[q] \longrightarrow 0.$$

Proof This follows from Lemma 7.1.

By Exercise 7.3, the group E_p/E_p^q is an $\mathbf{F}_q$-vector space of dimension $(p-1)/2$. If q does not divide $\#G = p-1$, Proposition 13.7 of chapter 13 implies that E_p/E_p^q is a *free* $\mathbf{F}_q[G^+]$-module of rank 1. This need not be true when q divides $p-1$. It does not hold for $p = 1129$ and $q = 3$, for instance [42].

It is difficult to say something general about the dimensions of the $\mathbf{F}_q$-vector spaces $Cl_p^+[q]$ and $Cl_p^-[q]$.

Proposition 7.5 *Let p be an odd prime. then*

$$h_p^- = 2p \prod_{\chi \text{ odd}} -\frac{1}{2} B_{1,\chi},$$

where χ runs over the odd characters of $G = \text{Gal}(\mathbf{Q}(\zeta_p)/\mathbf{Q})$. The generalized Bernoulli numbers $B_{1,\chi}$ are defined by $B_{1,\chi} = \sum_{a=1}^{p-1} \frac{a}{p} \chi(a)$.

Proof This classical result goes back to E.E. Kummer [23]. Since we do not need the result for the proof of Catalan's conjecture, we merely sketch the proof. See [50, Thm. 4.17] for more details. Briefly, the zeta function of $\mathbf{Q}(\zeta_p)$ is a product of the L-series $L(s, \chi)$, where χ runs over the characters $\chi : G \longrightarrow \mathbf{C}^*$. Similarly, the zeta function of the subfield $\mathbf{Q}(\zeta_p^+)$ is a product of the L-series $L(s, \chi)$, where χ runs over the characters of G^+ or, equivalently, over the *even* characters of G. Here a character χ is called *even* if $\chi(\iota) = 1$. It is called *odd* otherwise.

It follows that the product $\prod_{\chi \text{ odd}} L(1, \chi)$ is equal to the quotient of the residues in $s = 1$ of the zeta functions of $\mathbf{Q}(\zeta_p)$ and $\mathbf{Q}(\zeta_p^+)$, respectively. The result now follows from the formula for the residue in $s = 1$ of the zeta function of a number field and the fact that for odd characters χ we have

$$L(1, \chi) = -\frac{\pi i}{\tau(\chi^{-1})} B_{1,\chi}.$$

Here $\tau(\chi^{-1})$ denotes the *Gaussian sum* $-\sum_{a=1}^{p-1} \chi(a)^{-1} e^{2\pi ia/p} \in \mathbf{C}$. This completes the proof of the proposition. For more details, see [50, Thm. 4.9].

Table 7.1 Minus class numbers h_p^- for $p < 100$

p	h_p^-	p	h_p^-	p	h_p^-	p	h_p^-
3	1	19	1	43	211	71	3,882,809
5	1	23	3	47	695	73	11,957,417
7	1	29	8	53	4,889	79	100,146,415
11	1	31	9	59	41,241	83	838,216,959
13	1	37	37	61	76,301	89	13,379,363,737
17	1	41	121	67	85,3513	97	411,322,824,001

By means of this proposition, it is easy to compute the order h_p^- of the group Cl_p^-. The entries in Table 7.1 were already known to Kummer in 1850. The numbers h_p^- become very large as p grows. In practice, it is also rather easy to compute the *structure* of the group Cl_p^-. For instance, for $p = 29$, the class group Cl_p^- is isomorphic to $(\mathbf{Z}/2\mathbf{Z})^3$, while for $p = 31$, it is cyclic of order 9. See [41].

At present, there are no good methods to compute the orders h_p^+ of the plus parts Cl_p^+ of the class groups of $\mathbf{Q}(\zeta_p)$. For $p < 71$, the groups Cl_p^+ are known to be trivial. Under assumption of the general Riemann hypothesis, this result can be extended to the primes $p \le 163$. However, the group Cl_p^+ has not been computed for a single prime $p > 163$. It seems that the groups Cl_p^+ are usually small and often trivial.

For any given prime q, however, it is not difficult to compute the order of $Cl_p^+[q]$. For instance, for $p = 877$ and $q = 7$, the group $Cl_p^+[q]$ is a two-dimensional vector space over $\mathbf{F}_7$. See [42] for more information on the groups Cl_p^+.

Exercises

7.1 Let ε be an integer in a number field F with the property that $|\phi(\varepsilon)| = 1$ for all embeddings $\phi : F \hookrightarrow \mathbf{C}$. Show that it is a root of unity.

7.2 Show that $\mathbf{Z}[\zeta_p]$ is the ring of integers of $\mathbf{Q}(\zeta_p)$. Show that the $\mathbf{Z}[\zeta_p]$-ideal $\mathfrak{p}$ generated by $1 - \zeta_p$ is a prime ideal of norm p. See [50, p. 11].

7.3 Let p be an odd prime and let E_p denote the group of p-units of $\mathbf{Q}(\zeta_p)$.

(a) Show that the natural sequence

$$0 \longrightarrow \mathbf{Z}[\zeta_p]^* \longrightarrow E_p \xrightarrow{\text{ord}_\mathfrak{p}} \mathbf{Z} \longrightarrow 0$$

is exact. Here $\text{ord}_\mathfrak{p}$ is the valuation associated to the prime ideal $\mathfrak{p} = (1-\zeta_p)$ of the ring $\mathbf{Z}[\zeta_p]$.

(b) Show that for every odd prime $q \ne p$, the $\mathbf{F}_q$-dimension of E_p/E_p^q is equal to $(p-1)/2$.

7.4 Using techniques from algebraic number theory [24, Ch.V, s. 4], show directly that the class numbers of $\mathbf{Q}(\zeta_p)$ are equal to 1 when $p = 3, 5$, or 7.

7.5 Let $p > 2$ be prime and let $\mu \subset \mathbf{Q}(\zeta_p)^*$ denote the subgroup of roots of unity.

(a) Consider the homomorphism

$$\ker(Cl_p^+ \longrightarrow Cl_p) \longrightarrow \mu/\mu^2$$

that maps the class of a $\mathbf{Q}(\zeta_p^+)$-ideal I for which the $\mathbf{Z}[\zeta_p]$-ideal generated by I is equal to (α), to $\alpha^{1-\iota}$. Show that it is well defined and injective.

(b) Show that the map $\mathbf{Z}[\zeta_p]^* \longrightarrow \mu$ given by $u \mapsto u^{1-\iota}$ is well defined. Show that its image is μ^2.

(c) Show that the natural map $Cl_p^+ \longrightarrow Cl_p$ is injective.

7.6 Let $p > 2$ be prime. Show that the image W of the natural map $Cl_p^+ \longrightarrow Cl_p$ is invariant under ι. Show that the quotient of the group of ι-invariants of Cl_p, by the subgroup W, is annihilated by 2.

7.7 Let p be a prime. Show that the kernel and cokernel of the natural map $Cl_p^{1-\iota} \longrightarrow Cl_p \longrightarrow Cl_p^-$ are finite 2-groups. Show the same for the natural map $Cl_p[1+\iota] \longrightarrow Cl_p \longrightarrow Cl_p^-$. Here $Cl_p[1+\iota]$ denotes the kernel of the norm map given by $x \mapsto x\iota(x)$ for all $x \in Cl_p$.

7.8 Let H_0 be the subgroup of the obstruction group H that is defined by $H_0 = \{\alpha \in H : \mathrm{ord}_p(\alpha) \equiv 0 \pmod{q}\}$. Compute the index $[H : H_0]$.

8
Small p or q

In this chapter, we present Mihăilescu's proof of Theorem IV mentioned in chapter 1. The proof is by a p-adic argument. Let p, q be distinct odd primes. In Chapter 7, we introduced the obstruction group

$$H = \left\{ \alpha \in \mathbf{Q}(\zeta_p)^* : \mathrm{ord}_{\mathfrak{r}}(\alpha) \equiv 0 \ (\mathrm{mod}\ q) \text{ for all primes } \mathfrak{r} \neq \mathfrak{p} \right\} / \mathbf{Q}(\zeta_p)^{*q}.$$

By Exercise 7.2, the number $\pi = \zeta_p - 1$ generates the unique prime divisor of p in the ring $\mathbf{Z}[\zeta_p]$. We have $(\pi)^{p-1} = (p)$. Recall that $\iota \in G = \mathrm{Gal}(\mathbf{Q}(\zeta_p)/\mathbf{Q})$ denotes complex conjugation.

In Chapter 7, we associated the element $x - \zeta_p$ of H to a nonzero solution x, y of Catalan's equation $x^p - y^q = 1$. In this chapter, we show by means of a π-adic argument that the *minus component* $(x - \zeta_p)^{1-\iota}$ of $x - \zeta_p$ is a *nontrivial* element of H. This is the content of Theorem 8.3. We show that it easily implies Theorem IV. A result similar to, but weaker than, Theorem 8.3 was obtained by Y. Bugeaud and G. Hanrot [6].

Recall that saying that the element $(x - \zeta_p)^{1-\iota}$ of $H^- \subset H$ is trivial means that $(x - \zeta_p)^{1-\iota} = \alpha^q$ for some $\alpha \in \mathbf{Q}(\zeta_p)^*$ or, equivalently,

$$\frac{x - \zeta_p}{x - \iota(\zeta_p)} = \alpha^q,$$

for some $\alpha \in \mathbf{Q}(\zeta_p)^*$. Since $\mathbf{Q}(\zeta_p)$ does not contain any qth roots of unity, the element α is unique.

Definition *Let $x, y \in \mathbf{Z}$ be nonzero integers satisfying $x^p - y^q = 1$. Suppose that $(x - \zeta_p)^{1-\iota} = \alpha^q$ for some $\alpha \in \mathbf{Q}(\zeta_p)^*$. Let $w \in \overline{\mathbf{Q}}$ be a qth root of $\frac{x-\zeta_p}{1-\zeta_p}$ and set $w' = w/\alpha$. Then w' is a qth root of $\frac{x-\iota(\zeta_p)}{1-\zeta_p}$. Set*

$$\eta = (w - w')^q.$$

The number η is associated to our nonzero solution of Catalan's equation. Note that it does not depend on the choice of the qth root of w. Even though η is eventually

R. Schoof, *Catalan's Conjecture*, DOI: 10.1007/978-1-84800-185-5_8,
© Springer-Verlag London Limited 2008

shown not to exist, the analysis of its arithmetic properties plays a central role in our proof of Theorem 8.3.

Proposition 8.1 *Let p, q be two distinct odd primes. Suppose that $x, y \in \mathbf{Z}$ are nonzero integers satisfying $x^p - y^q = 1$ and that $(x - \zeta_p)^{1-\iota} = \alpha^q$ for some $\alpha \in \mathbf{Q}(\zeta_p)^*$. Then*

  (i) *we have $\alpha \equiv -1 \pmod{\pi}$,*
  (ii) *the number η defined above is a unit of the ring $\mathbf{Z}[\zeta_p]$,*
  (iii) *the norm from $\mathbf{Q}(\zeta_p)$ to $\mathbf{Q}$ of η is equal to 1.*

Proof (i) By Corollary 6.5, we have $x \equiv 1 \pmod{p}$. It follows that π divides $x - \zeta_p$. Moreover, the quotient $t = \frac{x - \zeta_p}{\pi} = \frac{x-1}{\pi} - 1$ is invertible modulo π. Since $\iota(\pi) = -1 + \zeta_p^{-1} = -\zeta_p^{-1}\pi$, we have

$$\alpha^q = t^{1-\iota}\frac{\pi}{-\zeta_p^{-1}\pi} \equiv -t^{1-\iota} \pmod{\pi}.$$

Since p is totally ramified in the extension $Q(\zeta_p)$ of $\mathbf{Q}$, the automorphism ι acts trivially modulo π. It follows that $t^{\iota-1} \equiv 1 \pmod{\pi}$ and hence α^q is congruent to $-1 \pmod{\pi}$.

On the other hand, multiplying the equation $(x - \zeta_p)^{1-\iota} = \alpha^q$ by its complex conjugate, we see that $(\alpha\iota(\alpha))^q = 1$. Since $\mathbf{Q}(\zeta_p)$ does not contain any nontrivial qth roots of unity, this implies that $\alpha\iota(\alpha) = 1$ and hence $\alpha^2 \equiv 1 \pmod{\pi}$. Combining this with the congruence $\alpha^q \equiv -1 \pmod{\pi}$, it follows that $\alpha \equiv -1 \pmod{\pi}$, as required.

  (ii) From the equality

$$\eta = (w - w')^q = w^q\left(1 - \frac{1}{\alpha}\right)^q$$

$$= \frac{x - \zeta_p}{1 - \zeta_p} \cdot \frac{x - \iota(\zeta_p)}{x - \zeta_p} \cdot (\alpha - 1)^q,$$

it follows that η is contained in $\mathbf{Q}(\zeta_p)$. By Corollary 6.5, the number x is congruent to 1 modulo p^{q-1}. This implies that $\frac{x-\zeta_p}{1-\zeta_p}$ is integral. Therefore, w is integral as well. Similarly, the fact that we have $w'^q = (w/\alpha)^q = \frac{x-\iota(\zeta_p)}{1-\zeta_p} = -\iota(w^q\zeta_p)$ implies that w' is integral. Therefore, η is integral and hence is in $\mathbf{Z}[\zeta_p]$. To see that it is a unit, we observe that $w - w'$ divides

$$w^q - w'^q = \frac{x - \zeta_p}{1 - \zeta_p} - \frac{x - \iota(\zeta_p)}{1 - \zeta_p} = \frac{-\zeta_p + \iota(\zeta_p)}{1 - \zeta_p} = \zeta_p^{-1} + 1,$$

which is a unit by Exercise 8.1. Therefore, $w - w'$ and $\eta = (w - w')^q$ are units as well.

(iii) Since η is a unit, the norm $N(\eta)$ of η, taken from $\mathbf{Q}(\zeta_p)$ to $\mathbf{Q}$, is equal to a unit in $\mathbf{Z}$. In other words, we have $N(\eta) = \pm 1$. We have the plus sign because the cyclotomic field $\mathbf{Q}(\zeta_p)$ is complex.

This proves the proposition.

Proposition 8.2 *Let p, q be two distinct odd primes. Suppose that $x, y \in \mathbf{Z}$ are nonzero integers satisfying $x^p - y^q = 1$ and suppose that $(x - \zeta_p)^{1-\iota} = \alpha^q$ for some $\alpha \in \mathbf{Q}(\zeta_p)^*$. Then we have*

$$q \equiv 1 \pmod{p}.$$

Proof We compute a π-adic approximation to the norm of η and confront this with the fact that $N(\eta) = 1$.

Since the degree of $\mathbf{Q}_p(\zeta_p)$ over the field of p-adic numbers $\mathbf{Q}_p$ is equal to $[\mathbf{Q}(\zeta_p) : \mathbf{Q}] = p - 1$, we can compute the norm of η *locally*, i.e., from $\mathbf{Q}_p(\zeta_p)$ to $\mathbf{Q}_p$. Let

$$\mu = \frac{x - 1}{1 - \zeta_p}.$$

The number μ is related to the numbers w and w'. We have

$$w^q = \frac{x - \zeta_p}{1 - \zeta_p} = 1 + \mu \quad \text{and} \quad w'^q = -\iota(\zeta_p)(1 + \iota(\mu)).$$

Recall that the unit $\eta = (w - w')^q$ does not depend on the choice of the qth root w of $\frac{x-\zeta_p}{1-\zeta_p}$. We choose $w \in \mathbf{Q}_p(\zeta_p)$ to be the value of the Taylor series for the qth root of $\frac{x-\zeta_p}{1-\zeta_p} = 1 + \mu$: we set

$$w = \sqrt[q]{1 + \mu} = \sum_{j=0}^{\infty} \binom{1/q}{j} \mu^j \text{ in } \mathbf{Q}_p(\zeta_p).$$

By Corollary 6.5, we have $x \equiv 1 \pmod{p^{q-1}}$, so that $\mu \equiv 0 \pmod{p^{q-2}}$. Exercise 8.2 implies therefore that the Taylor series does indeed converge to a number in $\mathbf{Q}_p(\zeta_p)$.

Claim *For $w' = w/\alpha$ in $\mathbf{Q}_p(\zeta_p)$ we have*

$$w' = -\zeta_p^r \sqrt[q]{1 + \iota(\mu)} = -\zeta_p^r \sum_{j=0}^{\infty} \binom{1/q}{j} \iota(\mu)^j.$$

Here $r \in \mathbf{Z}$ satisfies $rq \equiv -1 \pmod{p}$.

Proof of the claim If $p \not\equiv 1 \pmod{q}$, there are no primitive qth roots of unity in $\mathbf{Q}_p(\zeta_p)$. Therefore, there is a *unique* $w' \in \mathbf{Q}_p(\zeta_p)$ satisfying $w'^q = -\iota(\zeta_p)(1 + \iota(\mu))$. It can only be the qth root given by the Taylor series.

To see that this is also true when $p \equiv 1 \pmod{q}$, we observe that $w' = w/\alpha \equiv -1 \pmod{\pi}$. The Taylor series expression for the qth root of the number $-\iota(\zeta_p)$ $(1 + \iota(\mu))$ is also congruent to $-1 \pmod{\pi}$. Since w' is equal to a qth root of unity ξ times the value of the Taylor series, we conclude that $\xi = 1$. It follows that w' is necessarily equal to the number given by the Taylor series. This proves the claim.

Next we use the Taylor series expression to compute the norm of η. The qth power of $u = w - w'$ is equal to η, so that $N(\eta) = N(u)^q$. We perform our computation in the p-adic ring $\mathbf{Z}_p[\zeta_p]$. The maximal ideal of this local ring is generated by π.

We do the calculation modulo $\mu^2 \mathbf{Z}_p[\zeta_p]$. Expanding the first few terms of the Taylor series for the qth roots, we find that

$$u = w - w' = \sqrt[q]{1 + \mu} + \zeta_p^r \sqrt[q]{1 + \iota(\mu)}$$

$$\equiv \left(1 + \frac{\mu}{q}\right) + \zeta_p^r \left(1 + \frac{\iota(\mu)}{q}\right) \pmod{\mu^2}$$

$$\equiv (1 + \zeta_p^r)\left(1 + \frac{x-1}{q} \frac{1 - \zeta_p^{r+1}}{(1 - \zeta_p)(1 + \zeta_p^r)}\right) \pmod{\mu^2}.$$

By Exercise 8.1, the number $\zeta_p^r + 1$ is a unit of the ring $\mathbf{Z}[\zeta_p]$. Therefore, its norm is equal to 1 and the norm of u is given by

$$N(u) \equiv \prod_{\substack{\zeta \in \mu_p \\ \zeta \neq 1}} \left(1 + \frac{x-1}{q} \frac{1 - \zeta^{r+1}}{(1 - \zeta)(1 + \zeta^r)}\right) \pmod{\mu^2}$$

$$\equiv 1 + \frac{x-1}{q} \sum_{\substack{\zeta \in \mu_p \\ \zeta \neq 1}} \frac{1 - \zeta^{1+r}}{(1 - \zeta)(1 + \zeta^r)} \pmod{\mu^2}.$$

The summand corresponding to ζ_p is equal to

$$\frac{1 - \zeta_p^{1+r}}{(1 - \zeta_p)(1 + \zeta_p^r)} = \frac{1 - (1 + \pi)^{r+1}}{-\pi(1 + (1 + \pi)^r)}$$

$$= \frac{-(r+1)\pi + O(\pi^2)}{-\pi(2 + r\pi) + O(\pi^3)}$$

$$\equiv \frac{r+1}{2} \pmod{\pi},$$

where $O(\pi^k)$ denotes a term that is divisible by π^k. Since the Galois group of $\mathbf{Q}_p(\zeta_p)$ over $\mathbf{Q}_p$ preserves the ideal generated by π, it follows that

$$\frac{1 - \zeta^{1+r}}{(1 - \zeta)(1 + \zeta^r)} \equiv \frac{r+1}{2} \pmod{\pi} \qquad \text{for every } \zeta \in \mu_p, \zeta \neq 1.$$

Since the number $\frac{r+1}{2}$ does not depend on ζ, we find that

$$N(u) \equiv 1 + \frac{x-1}{q}\frac{(r+1)(p-1)}{2} \pmod{(x-1)\pi}.$$

Here we use that $\pi(x-1)$ divides $\mu^2 = ((x-1)/\pi)^2$, which in turn follows from the fact that $x \equiv 1 \pmod{p^{q-1}}$. Since $rq \equiv -1 \pmod{p}$, we have

$$N(\eta) = N(u)^q \equiv 1 + (x-1)\frac{1-q}{2q} \pmod{(x-1)\pi}.$$

Since $N(\eta) = 1$, this implies that π and hence p divide $1-q$.

This proves the proposition.

Theorem 8.3 *Suppose that p, q are distinct odd primes and that x, y are nonzero integers satisfying Catalan's equation $x^p - y^q = 1$. Then the minus component $(x - \zeta_p)^{1-\iota}$ of $x - \zeta_p$ is a nontrivial element in the obstruction group H.*

Proof Suppose that $(x - \zeta_p)^{1-\iota}$ is trivial in H. Then we have

$$\frac{x - \zeta_p}{x - \iota(\zeta_p)} = \alpha^q$$

for some $\alpha \in \mathbf{Q}(\zeta_p)^*$. By Proposition 8.2, we have $q \equiv 1 \pmod{p}$.

We consider once again the unit

$$u = \sqrt[q]{1 + \mu} + \zeta_p^r \sqrt[q]{1 + \iota(\mu)}$$

that was defined above. We compute the norm of $\eta = u^q$ using Taylor series expansions. This time we do our calculation modulo μ^3. Since $q \equiv 1 \pmod{p}$ and $rq \equiv -1 \pmod{p}$, we have $r \equiv -1 \pmod{p}$ and we see that the linear term $\mu + \zeta_p^r \iota(\mu)$ in the Taylor series expansion of u vanishes.

We compute the degree 2 terms of the Taylor series of the qth roots. Recall that $\mu = -\frac{x-1}{\pi}$, so that $\iota(\mu) = \zeta_p \mu$. We find

$$
\begin{aligned}
u = w - w' &= \sqrt[q]{1 + \mu} + \zeta_p^r \sqrt[q]{1 + \iota(\mu)} \\
&\equiv 1 + \binom{1/q}{2}\mu^2 + \zeta_p^{-1}\left(1 + \binom{1/q}{2}\iota(\mu)^2\right) \pmod{\mu^3} \\
&\equiv (1 + \zeta_p^{-1})\left(1 + \frac{1-q}{2q^2}(x-1)^2\frac{\zeta_p}{(1-\zeta_p)^2}\right) \pmod{\mu^3}.
\end{aligned}
$$

The norm of u is equal to

$$N(u) = 1 + \frac{1-q}{2q^2}(x-1)^2 \sum_{\substack{\zeta \in \mu_p \\ \zeta \neq 1}} \frac{\zeta}{(1-\zeta)^2} \pmod{\mu^3}.$$

By Exercise 8.2, the sum of the series is equal to $\frac{1-p^2}{12}$. It follows that

$$N(\eta) = N(u)^q \equiv 1 + \frac{1-q}{2q}(x-1)^2 \frac{1-p^2}{12} \pmod{\mu^3}.$$

Note that $(x-1)^2/12$ is a π-adic integer.

Since $N(\eta) = 1$, the number $\mu^3 = ((x-1)/\pi)^3$ divides $\frac{1-q}{2q}(x-1)^2 \frac{1-p^2}{12}$. This implies that $x-1$ divides $\frac{1-q}{2q}\frac{1-p^2}{12}\pi^3$ in $\mathbf{Z}_p[\zeta_p]$. Since Corollary 6.5 says that $x \equiv 1 \pmod{p^{q-1}}$, it follows that p^{q-1} divides $(1-q)\frac{\pi^3}{3}$. Since $\pi^3/3$ divides p, it follows that p^{q-1} divides $(q-1)p$ and hence that p^{q-2} divides $q-1$. This is impossible since p^{q-2} is larger than $q-1$.

This contradiction proves the theorem.

Corollary 8.4 *Let p, q be odd primes and suppose that q does not divide h_p^- or that p does not divide h_q^-. Then Catalan's equation $x^p - y^q = 1$ has no nonzero solutions $x, y \in \mathbf{Z}$.*

Proof By symmetry, it suffices to consider the case where q does not divide h_p^-. By the remarks at the end of Section 7, we have $H^- = Cl^-[q]$, so that the minus part H^- of the obstruction group H is trivial. However, by Theorem 8.3, a nonzero solution to Catalan's equation gives rise to the *nontrivial* element $(x-\zeta_p)^{1-\iota}$ in H^-. It follows that there are no nonzero solutions to Catalan's equation.

Theorem IV of Chapter 1 is an immediate consequence:

Theorem IV *If p, q are odd primes and either p or q does not exceed 41, then Catalan's equation $x^p - y^q = 1$ admits no nonzero solutions $x, y \in \mathbf{Z}$.*

Proof By Lemma 6.1, we have $p \neq q$. By symmetry, we may assume that $p < q$. From the table of minus class numbers (Table 7.1), we see the following. When $p \leq 19$, the minus class number h_p^- is equal to 1, so that q does not divide h_p^- for any prime q. For these primes, the theorem is clear. An inspection of Table 7.1 furthermore shows that for $p = 23, 29, 31, 37$, and 41, the minus class number h_p^- is equal to $3, 8, 3^2, 37$, and 11^2, respectively. In each case, the prime divisors of h_p^- are at most p and once again q, being larger than p, cannot divide h_p^-. Corollary 8.4 therefore implies the theorem.

With some more effort, one can extend Theorem IV to $p = 43$. See Exercise 8.5. However, for $p = 47$, this approach fails. The minus class numbers of $\mathbf{Q}(\zeta_{47})$ and $\mathbf{Q}(\zeta_{139})$ are given by

$$h_{47}^- = 5 \cdot 139,$$
$$h_{139}^- = 3^2 \cdot 47 \cdot 277^3 \cdot 967 \cdot 1188961909.$$

It follows that when $q = 139$, we have that p divides h_q^- as well as q divides h_p^-.

For the purpose of proving Theorems II and III in Chapters 11 and 12, we only need to know that Catalan's equation $x^p - y^q = 1$ has no solution in nonzero

$x, y \in \mathbf{Z}$ when one of p, q is at most 5. In order to deduce this from Corollary 8.4, it suffices to know that one has $h_p^- = 1$ for $p = 3$ and 5. This can easily be established with standard techniques from algebraic number theory. See Exercise 7.4.

Exercises

8.1 Let p be an odd prime. Show that $1 + \zeta_p$ is a unit of the ring $\mathbf{Z}[\zeta_p]$.

8.2 Let $\alpha \in \mathbf{Q}$.

(a) Show that the series $\sum_{k \geq 0} \binom{\alpha}{k} t^k$ converges for $t \in \mathbf{R}$ with $|t| < 1$. Show that the sum is equal to $(1 + t)^\alpha$.

(b) Let p be a prime number and suppose that $\alpha = r/s$ with $r, s \in \mathbf{Z}$ and $s \not\equiv 0 \pmod{p}$. Show that the series $\sum_{k \geq 0} \binom{\alpha}{k} t^k$ converges for $t \in p\mathbf{Z}_p$ and show that the sth power of the sum is equal to $(1 + t)^r$.

8.3 Check that the double Wieferich criterion of Chapter 10 holds for the pair $p = 47$ and $q = 139$ and conclude that Catalan's equation has no nonzero solutions for these exponents.

8.4 Let n be a natural number and let μ_n denote the group of nth roots of unity in $\mathbf{C}^*$. For every $k \geq 1$, we let $s_k = \sum\limits_{\substack{\zeta \in \mu_n \\ \zeta \neq 1}} \frac{1}{(1-\zeta)^k}$. We have $s_k \in \mathbf{Q}$.

(a) Show the following equalities of power series in $\mathbf{Q}[[X]]$:

$$\sum_{k \geq 1} s_k \frac{X^k}{k} = -\log\left(\prod_{\substack{\zeta \in \mu_n \\ \zeta \neq 1}} (1 - \frac{X}{1 - \zeta})\right)$$

$$= -\log\left(\frac{(1 - X)^n - 1}{-nX}\right)$$

$$= \frac{n-1}{2} X - \frac{(n-1)(n-5)}{24} X^2 + O(X^3).$$

(b) Show that one has

$$\sum_{\substack{\zeta \in \mu_n \\ \zeta \neq 1}} \frac{\zeta}{(1 - \zeta)^2} = s_2 - s_1 = \frac{1 - n^2}{12}.$$

8.5 Extend Theorem IV to $p = 43$:

(a) show that $h_{43}^- = 211$;

(b) show that

$$h_{211}^- = 3^2 \cdot 7^2 \cdot 41 \cdot 71 \cdot 181 \cdot 281^2 \cdot 421 \cdot 1051 \cdot 12251 \cdot 113981701 \cdot 4343510221,$$

and conclude that the hypothesis of Corollary 8.4 still holds.

9
The Stickelberger Ideal

Let p be an odd prime and let ζ_p denote a primitive pth root of unity. Let $G = \mathrm{Gal}(\mathbf{Q}(\zeta_p)/\mathbf{Q})$. We have $G = \{\sigma_a : a \in (\mathbf{Z}/p\mathbf{Z})^*\}$, where σ_a is defined by $\sigma_a(\zeta_p) = \zeta_p^a$. The Stickelberger ideal is an ideal of the group ring $\mathbf{Z}[G]$. In this chapter, we establish some of its basic properties. See also [25] and [50]. We determine the $\mathbf{Z}$-rank of the Stickelberger ideal. This is the content of Theorem 9.3. The most interesting property of the Stickelberger ideal is that it annihilates the ideal class group of $\mathbf{Q}(\zeta_p)$. This is the content of Theorem 9.6.

Definition *The Stickelberger ideal is the ideal of $\mathbf{Z}[G]$ generated by the elements*

$$\theta_i = \sum_{a=1}^{p-1} \left[\frac{ia}{p}\right] \sigma_a^{-1}, \quad i \in \mathbf{Z}.$$

Recall that $[x]$ denotes the integral part of $x \in \mathbf{R}$. The *Stickelberger element* is $\theta_p = \sum_{a=1}^{p-1} a\sigma_a^{-1}$. The *G-trace* is the element $T = \sum_{a=1}^{p-1} \sigma_a$ of $\mathbf{Z}[G]$.

Lemma 9.1 *Let p be an odd prime and let $G = \mathrm{Gal}(\mathbf{Q}(\zeta_p)/\mathbf{Q})$. Then the following relations hold in the group ring $\mathbf{Z}[G]$:*

(i) $\theta_{p+i} = \theta_i + \theta_p$ for every $i \in \mathbf{Z}$;
(ii) $\frac{1}{p}(\sigma_i - i)\theta_p = -\theta_i$ for every $i \not\equiv 0 \pmod{p}$;
(iii) $\theta_i + \theta_{p-i} = \theta_p - T$ for every $i \not\equiv 0 \pmod{p}$;
(iv) for every $j \not\equiv 0 \pmod{p}$, we have

$$\sigma_j\theta_p = -p\theta_j + j\theta_p,$$
$$\sigma_j\theta_i = \theta_{ij} - i\theta_j \quad \text{for all } i \not\equiv 0 \pmod{p}.$$

Proof Parts *(i)* and *(ii)* are immediate from the definition of θ_i. Note that for every $i \not\equiv 0 \pmod{p}$, the coefficients of $(\sigma_i - i)\theta_p$ are divisible by p. For *(iii)*, note that

$$\theta_i + \theta_{p-i} = \sum_{a=1}^{p-1}\left(\left[\frac{ai}{p}\right] + \left[\frac{(p-i)a}{p}\right]\right)\sigma_a^{-1} = \sum_{a=1}^{p-1}(a-1)\sigma_a^{-1} = \theta_p - T$$

for every i prime to p. Part *(iv)* is also easily verified.

R. Schoof, *Catalan's Conjecture*, DOI: 10.1007/978-1-84800-185-5_9,
© Springer-Verlag London Limited 2008

Definition *Using the notation of Lemma 9.1, we define for $i \in \mathbf{Z}$ the following elements in the ring $\mathbf{Z}[G]$. Recall that $\iota = \sigma_{-1}$.*

$$f_i = \theta_{i+1} - \theta_i,$$
$$e_i = (1 - \iota)f_i.$$

Proposition 9.2 *As a group, the Stickelberger ideal is generated by the G-trace T and by the elements f_i for $1 \leq i \leq (p-1)/2$.*

Proof Let J be the subgroup of $\mathbf{Z}[G]$ generated by the elements $f_1, \ldots, f_{\frac{p-1}{2}}$ and by the G-trace T. Since $\theta_1 = 0$, the group J is also generated by T and by the elements $\theta_2, \ldots, \theta_{\frac{p+1}{2}}$. From Lemma 9.1 *(iii)*, it follows that T and hence all of J is contained in the Stickelberger ideal. Conversely, it suffices to show that the elements in J generate the Stickelberger ideal as a group. Taking $i = \frac{p+1}{2}$ in the relation of Lemma 9.1 *(iii)*, we see that $\theta_p - T \in J$. This implies that $\theta_p \in J$. The same formula implies that $\theta_i \in J$ for $\frac{p+1}{2} < i \leq p - 1$. It follows from Lemma 9.1 (iii) that J contains the elements $\theta_1, \theta_2, \theta_3, \ldots, \theta_p$. The periodicity relation of Lemma 9.1 *(i)* implies then that we have $\theta_i \in J$ for every $i \in \mathbf{Z}$. Since the θ_i generate the Stickelberger ideal over $\mathbf{Z}[G]$, it suffices now to show that J itself is a $\mathbf{Z}[G]$-ideal. This follows at once from the relations in Lemma 9.1 *(iv)*.

Theorem 9.3 *Let p be an odd prime number and $G = \mathrm{Gal}(\mathbf{Q}(\zeta_p)/\mathbf{Q})$.*

(i) *The elements f_i, for $1 \leq i \leq \frac{p-1}{2}$, together with the G-trace T form a $\mathbf{Z}$-basis for the Stickelberger ideal of $\mathbf{Z}[G]$.*
(ii) *Let I be the ideal of $\mathbf{Z}[G]$ that is the product of the Stickelberger ideal by $(1-\iota)$. The elements e_i, for $1 \leq i \leq \frac{p-1}{2}$, form a $\mathbf{Z}$-basis for I.*

Proof By Proposition 9.2, the elements f_i, for $1 \leq i \leq \frac{p-1}{2}$, together with the G-trace T generate the Stickelberger ideal as a group. In order to show that the generators are *independent*, it suffices to show that the $\mathbf{Z}$-rank of the Stickelberger ideal is equal to the number of generators $\frac{p+1}{2}$. Similarly, since $(1 - \iota)T = 0$, the elements $e_i = (1 - \iota)f_i$, for $1 \leq i \leq \frac{p-1}{2}$, generate I as a group and, in order to prove *(ii)*, it suffices to show that the $\mathbf{Z}$-rank of I is equal to $\frac{p-1}{2}$. The $\mathbf{Z}$-ranks of these ideals are equal to the $\mathbf{C}$-dimension of the ideals generated by the same elements in the group ring $\mathbf{C}[G]$. We tensor the group ring therefore with $\mathbf{C}$. It follows from the formulas of Lemma 9.1 *(i)* and *(ii)* that the ideal generated by the Stickelberger ideal inside $\mathbf{C}[G]$ is generated by the Stickelberger element θ_p. Similarly, the ideal generated by I inside $\mathbf{C}[G]$ is generated by $(1 - \iota)\theta_p$.

For every character $\chi : G \longrightarrow \mathbf{C}^*$, we let $\chi : \mathbf{C}[G] \longrightarrow \mathbf{C}$ denote the $\mathbf{C}$-algebra homomorphism induced by χ. In other words, for $\theta = \Sigma_{\sigma \in G} x_\sigma \sigma \in \mathbf{C}[G]$, we set $\chi(\theta) = \Sigma_{\sigma \in G} x_\sigma \chi(\sigma)$. Taking the product over all characters χ, we obtain the $\mathbf{C}$-algebra isomorphism

$$j : \mathbf{C}[G] \quad \overset{\cong}{\longrightarrow} \quad \prod_{\chi} \mathbf{C},$$

which is given by $j(\theta) = (\chi(\theta))_\chi$.

Consider the element $\theta_p = \Sigma_{a=1}^{p-1} a\sigma_a^{-1}$ of the Stickelberger ideal. The isomorphism j maps it to the vector $\left(\chi(\theta_p)\right)_\chi$. When χ is an *odd* character, i.e., when $\chi(\iota) = -1$, we have

$$\chi(\theta_p) = \sum_{a=1}^{p-1} a\chi(\sigma_a)^{-1} = pB_{1,\chi}.$$

Here $B_{1,\chi}$ is the generalized Bernoulli number that is defined in chapter 7. By Proposition 7.5, the product over the odd characters χ of $-\frac{1}{2}B_{1,\chi}$ is equal to $h_p^-/2p$. In particular, it is not zero. It follows that $B_{1,\chi}$ and hence $\chi(\theta_p)$ are not zero for each of the $\frac{p-1}{2}$ odd characters χ.

When χ is even, we have that

$$\chi(\theta_p) = \sum_{a=1}^{(p-1)/2} (a + (p-a))\chi(\sigma_a)^{-1} = p \sum_{a=1}^{(p-1)/2} \chi(\sigma_a)^{-1}.$$

The latter expression is 0 when $\chi \neq 1$, while it is equal to $p(p-1)/2$ for $\chi = 1$.

It follows that $\chi(\theta_p) \neq 0$ for precisely $\frac{p+1}{2}$ different characters of G. Therefore, the $\mathbf{C}[G]$-ideal generated by the Stickelberger ideal inside $\mathbf{C}[G]$ has dimension $\frac{p+1}{2}$. This implies that the $\frac{p+1}{2}$ elements of the Stickelberger ideal are independent, as required.

Similarly, since $\chi(1 - \iota)$ is zero if χ is an even character of the Galois group G, while it is equal to 2 when χ is odd, we have $\chi((1 - \iota)\theta_p) \neq 0$ precisely for the $\frac{p-1}{2}$ odd characters of G. It follows that the $\mathbf{C}[G]$-ideal generated by the $\mathbf{Z}[G]$-ideal I inside $\mathbf{C}[G]$ has dimension $\frac{p-1}{2}$. This implies that the $\frac{p-1}{2}$ elements e_i of I are independent, as required.

This proves the theorem.

For $a, i \in \mathbf{Z}$, we set $n_{i,a} = [\frac{ia}{p}]$ and $m_{i,a} = n_{i+1,a} - n_{i,a}$. Then we have, for $i \in \mathbf{Z}$,

$$\theta_i = \sum_{a=1}^{p-1} n_{i,a}\sigma_a^{-1},$$

$$f_i = \theta_{i+1} - \theta_i = \sum_{a=1}^{p-1} m_{i,a}\sigma_a^{-1}.$$

For $a, i \in \mathbf{Z}$, we define the coefficients $u_{i,a}$ by

$$e_i = (1 - \iota)f_i = \sum_{a=1}^{p-1} u_{i,a}\sigma_a^{-1}.$$

Proposition 9.4 *Using the notation introduced above, exactly* $(p-1)/2$ *of the coefficients* u_{ia} *with* $1 \le a \le p-1$, *are equal to 1 while the other* $(p-1)/2$ *are equal to* -1. *There is a symmetry given by* $e_{p-1-i} = e_i$ *for* $1 \le i \le (p-1)/2$

Proof Since $m_{i,a} = \left[\frac{(i+1)a}{p}\right] - \left[\frac{ia}{p}\right]$, it follows from Exercise 9.3 that the coefficients $m_{i,a}$ are either 0 or 1. By Exercise 9.4, the element $(1 + \iota)f_i$ is equal to the G-trace $T = \Sigma_{\sigma \in G}\sigma$. Therefore, the sum $m_{i,a} + m_{i,p-a}$ is always equal to 1. It follows that exactly one of $m_{i,a}$, $m_{i,p-a}$ is zero, while the other is 1. In particular, exactly $(p-1)/2$ of the coefficients $m_{i,a}$ are 0 and the other $(p-1)/2$ are equal to 1. Since we have $e_i = (1 - \iota)f_i$ for $1 \le i \le \frac{p-1}{2}$, we see that exactly one of the coefficients $u_{i,a}$, $u_{i,p-a}$ is 1, while the other is equal to -1.

Finally, it is easy to see that there is a symmetry given by $f_{p-1-i} = f_i$ for $0 \le i \le (p-1)/2$, and, therefore, there is a symmetry given by $e_{p-1-i} = e_i$. This proves the proposition.

We give a numerical example. Let $p = 7$. Tables 9.1–9.3 list the elements θ_i, f_i, and e_i for $i = 1, \ldots, 6$. Recall that $f_{6-i} = f_i$ and $e_{6-i} = e_i$ in this range.

Table 9.1 Elements $\theta_i = \sum_{a=1}^{p-1} n_{i,a}\sigma_a^{-1}$

i	$n_{i,1}$	$n_{i,2}$	$n_{i,3}$	$n_{i,4}$	$n_{i,5}$	$n_{i,6}$
1	0	0	0	0	0	0
2	0	0	0	1	1	1
3	0	0	1	1	2	2
4	0	1	1	2	2	3
5	0	1	2	2	3	4
6	0	1	2	3	4	5

Table 9.2 Elements $f_i = \theta_{i+1} - \theta_i = \sum_{a=1}^{p-1} m_{i,a}\sigma_a^{-1}$

i	$m_{i,1}$	$m_{i,2}$	$m_{i,3}$	$m_{i,4}$	$m_{i,5}$	$m_{i,6}$
1	0	0	0	1	1	1
2	0	0	1	0	1	1
3	0	1	0	1	0	1

Table 9.3 Elements $e_i = (1 - \iota)f_i = \sum_{a=1}^{p-1} u_{i,a}\sigma_a^{-1}$

i	$u_{i,1}$	$u_{i,2}$	$u_{i,3}$	$u_{i,4}$	$u_{i,5}$	$u_{i,6}$
1	-1	-1	-1	1	1	1
2	-1	-1	1	-1	1	1
3	-1	1	-1	1	-1	1

The following theorem is the key ingredient in the proof of Stickelberger's theorem. In his 1890 paper, L. Stickelberger [47] proved his result for arbitrary cyclotomic fields. We prove it here only for the cyclotomic fields $\mathbf{Q}(\zeta_p)$. In that case, the result was already obtained by E.E. Kummer [23] in 1847.

Fig. 9.1 Ernst Eduard Kummer (1810–1893)

Theorem 9.5 *Let p be a prime and let $\mathfrak{l}$ be a prime ideal of the ring $\mathbf{Z}[\zeta_p]$ of degree 1. Then $\mathfrak{l}^\theta$ is principal for every element θ of the Stickelberger ideal.*

Proof Set $F = \mathbf{Q}(\zeta_p)$ and let $G = \mathrm{Gal}(F/\mathbf{Q})$. As before, we write $G = \{\sigma_a : a \in (\mathbf{Z}/p\mathbf{Z})^*\}$, where $\sigma_a(\zeta_p) = \zeta_p^a$. Let $\mathfrak{l}$ be a prime ideal of $\mathbf{Z}[\zeta_p]$ of degree 1. Its residue field is $\mathbf{F}_l$, where l is a prime congruent to 1 (mod p). Set $\Delta = \mathrm{Gal}(\mathbf{Q}(\zeta_l)/\mathbf{Q})$. The restriction maps induce an isomorphism $\mathrm{Gal}(F(\zeta_l)/\mathbf{Q}) \longrightarrow G \times \Delta$

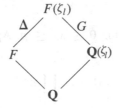

We let G act on $F(\zeta_l)$ by setting $\sigma_a(\zeta_l) = \zeta_l$. Similarly, we write

$$\Delta = \{\rho_c : c \in (\mathbf{Z}/l\mathbf{Z})^*\},$$

where $\rho_c(\zeta_l) = \zeta_l^c$, and we extend the action of Δ to $F(\zeta_l)$ by setting $\rho_c(\zeta_p) = \zeta_p^c$. We have $G = \mathrm{Gal}(F(\zeta_l)/\mathbf{Q}(\zeta_l))$ and $\Delta = \mathrm{Gal}(F(\zeta_l)/F)$.

We fix a primitive lth root of unity ζ_l and a nontrivial character χ from $(\mathbf{Z}/l\mathbf{Z})^*$ to the group of pth roots of unity μ_p in $\mathbf{Z}[\zeta_p]^*$. Then we define the *Gaussian sum* τ by

$$\tau = - \sum_{x(\mathrm{mod}\ l)} \chi(x)\zeta_l^x.$$

The Gaussian sum τ is contained in the ring $\mathbf{Z}[\zeta_p, \zeta_l]$. By Exercise 9.6, we have

$$\rho_c(\tau) = \chi^{-1}(c)\tau \quad \text{for each } \rho_c \in \Delta.$$

This shows that τ^p is Δ-invariant and is hence contained in the p-th cyclotomic field F. The subfield $F(\tau)$ of $F(\zeta_l)$ is a *Kummer extension* of F. By Exercise 9.9, the Kummer homomorphism

$$\mathrm{Gal}(F(\tau)/F) \quad \longrightarrow \quad \mathrm{Hom}(\langle \tau^p \rangle F^{*p}/F^{*p}, \mu_p)$$

given by $\sigma \mapsto \varphi$, where φ is determined by $\varphi(\tau) = \sigma(\tau)/\tau$, is an isomorphism of G-modules. Here G acts on $\mathrm{Gal}(F(\tau)/F)$ by conjugation and acts on the right-hand group via $\sigma_a(\varphi)(\tau) = \sigma_a(\varphi(\sigma_a^{-1}(\tau)))$ for every $\sigma_a \in G$.

Since τ is contained in the cyclotomic field $F(\zeta_l)$, the extension $\mathbf{Q} \subset F(\tau)$ is *abelian* and the action of G on $\mathrm{Gal}(F(\tau)/F)$ is *trivial*. It follows that the action by G on $\mathrm{Hom}(\langle \tau^p \rangle F^{*p}/F^{*p}, \mu_p)$ is also trivial. This means that G acts on $\langle \tau^p \rangle F^{*p}/F^{*p}$ the same way it acts on μ_p. In other words, we have

$$\sigma_b(\tau^p) = \tau^{pb} \cdot (p\text{th power})$$

for all $b \in (\mathbf{Z}/p\mathbf{Z})^*$.

By Exercise 9.6, the Gaussian sum τ is an algebraic integer that divides l. Therefore, we have the following factorization of the F-ideal generated by τ^p:

$$(\tau^p) = \prod_{\sigma_a \in G} \sigma_a^{-1}(l)^{x_a}$$

for certain integers x_a satisfying $0 \leq x_a \leq p$. Applying the automorphism σ_b, we find

$$\sigma_b(\tau^p) = \prod_{\sigma_a \in G} \sigma_{ba^{-1}}(l)^{x_a} = \prod_{\sigma_a \in G} \sigma_a^{-1}(l)^{bx_a} \cdot (p\text{th power}).$$

Replacing a by ab in the product on the left, we obtain

$$x_{ba} \equiv bx_a \pmod{p}, \quad \text{for all } a, b \in (\mathbf{Z}/p\mathbf{Z})^*.$$

Setting $a = 1$, this gives $x_b \equiv bx_1 \pmod{p}$ for all $b \in (\mathbf{Z}/p\mathbf{Z})^*$. We claim that $x_1 \not\equiv 0 \pmod{p}$. Indeed, if $x_1 \equiv 0 \pmod{p}$, then $x_b \equiv 0 \pmod{p}$ for every $b \in (\mathbf{Z}/p\mathbf{Z})^*$. This implies that the F-ideal generated by τ^p is a pth power. It follows from Exercise 9.8 that the extension $F \subset F(\tau)$ is unramified at l. This cannot be true, since $F(\tau)$ is a degree p subfield of the totally ramified extension $F(\zeta_l)$ of F. It follows that x_1 is invertible modulo p and that the set $\{x_a : a \in (\mathbf{Z}/p\mathbf{Z})^*\}$ is equal to $\{1, 2, \ldots, p-1\}$. In particular, there is an element $a \in (\mathbf{Z}/p\mathbf{Z})^*$ for which $x_a = 1$. We have $x_{ba} = b$ for all $b = 1, 2, \ldots, p-1$ and hence

$$\sigma_a(\tau^p) = \prod_{\sigma_b \in G} \sigma_a \sigma_b^{-1}(\mathfrak{l})^{x_b} = \prod_{\sigma_b \in G} \sigma_b^{-1}(\mathfrak{l})^{x_{ab}} = \prod_{\sigma_b \in G} \sigma_b^{-1}(\mathfrak{l})^b.$$

We find that

$$\mathfrak{l}^{\theta_p} = \sigma_a(\tau^p),$$

where $\theta_p \in \mathbf{Z}[G]$ is the Stickelberger element $\theta_p = \Sigma_{b=1}^{p-1} b\sigma_b^{-1}$. In particular, we see that the ideal $\mathfrak{l}^{\theta_p}$ is principal.

Applying the elements $\sigma_i - i \in \mathbf{Z}[G]$, we find by Lemma 9.1 (ii) that

$$\mathfrak{l}^{p\theta_i} = \sigma_a(\tau^{(i-\sigma_i)p}).$$

We claim that the element $\tau^{i-\sigma_i}$ is contained in F. Indeed, for each $c \in (\mathbf{Z}/l\mathbf{Z})^*$, we have

$$\tau^{(\sigma_i-i)(\rho_c-1)} = \tau^{(\rho_c-1)(\sigma_i-i)} = 1.$$

This follows from the fact that τ^{ρ_c-1} is contained in the group μ_p of the pth roots of unity and that the $\mathbf{Z}[G]$-module μ_p is annihilated by $\sigma_i - i$ for all $i \not\equiv 0 \pmod{p}$.

It follows that this is an equality of the pth powers of F-ideals. Since the ideal group is torsion-free, we obtain the relation

$$\mathfrak{l}^{\theta_i} = \sigma_a(\tau^{i-\sigma_i}).$$

Therefore, $\mathfrak{l}^{\theta}$ is principal for every element θ of the Stickelberger ideal, as required.

The proof of Stickelberger's theorem is concluded by showing that the ideal class group is *generated* by the classes of the prime ideals of degree 1. The usual proof to show this relies on a density theorem and on class field theory. We present a different, more elementary proof due to E.E. Kummer [23]. See also [28].

Theorem 9.6 *Let p be prime number. Then the Stickelberger ideal annihilates the ideal class group of $\mathbf{Q}(\zeta_p)$.*

Proof By Theorem 9.5, it suffices to show that the ideal class group of $\mathbf{Q}(\zeta_p)$ is generated by the classes of the primes of degree 1. Thus, let $\mathfrak{q}$ be a prime ideal of $\mathbf{Z}[\zeta_p]$ of degree $f \geq 2$. By Kummer's Lemma (Exercise 9.7), we have $\mathfrak{q} = (q, h(\zeta_p))$ for some prime number $q \not\equiv 1 \pmod p$ and some irreducible monic divisor $h(X)$ of the cyclotomic polynomial $\Phi_p(X) \in \mathbf{F}_q[X]$. The degree of h is f and we have

$$h(X) = \prod_{j=1}^{f}(X - \xi^{q^j})$$

for some primitive pth root of unity ξ in $\overline{\mathbf{F}}_q$. It follows that the constant term of H is equal to $(-1)^f$ times a pth root of unity. Since $f \geq 2$, the only pth root of unity contained in $\mathbf{F}_q$ is 1. Therefore,

$$h(0) = (-1)^f.$$

Let $H(X) \in \mathbf{Z}[X]$ denote a monic lift of $h(X)$ with the property that $H(0) = (-1)^f$. For $j = 1, 2$, we analyze the prime divisors of the elements $H(\zeta_p) + q^j$ of $\mathbf{Z}[\zeta_p]$. Let $\mathfrak{r}$ be a prime ideal of degree at least 2. Then it is of the form $\mathfrak{r} = (r, \tilde{h}(\zeta_p))$ for some prime number $r \not\equiv 1 \pmod p$ and some irreducible monic divisor $h'(X)$ of the cyclotomic polynomial $\Phi_p(X) \in \mathbf{F}_r[X]$. Let $\tilde{f}$ denote the degree of $\tilde{h}(X)$. The constant term of $\tilde{h}(X)$ is equal to $(-1)^{\tilde{f}}$.

The prime $\mathfrak{r}$ divides $H(\zeta_p) + q^j$ if and only if the polynomial $\tilde{h}(X)$ divides $H(X) + q^j$ in the ring $\mathbf{F}_r[X]$. This shows that necessarily $\tilde{f} \leq f$. If $\tilde{f} = f$, the fact that both polynomials are monic implies that the polynomials $\tilde{h}(X)$ and $H(X) + q^j$ are *equal* in the ring $\mathbf{F}_r[X]$. Inspection of their constant terms shows that r divides $q^j + (-1)^f - (-1)^{\tilde{f}} = q^j$. This implies that $r = q$ and hence $\mathfrak{r} = \mathfrak{q}$. Finally, we observe that $\mathfrak{q}$ divides either $H(\zeta_p) + q$ or $H(\zeta_p) + q^2$ exactly once. Indeed, if this were not true, then $\mathfrak{q}^2$ would divide the difference $q^2 - q$. This is impossible, because q is unramified in $\mathbf{Q}(\zeta_p)$.

We conclude that $\tilde{f} < f$ and hence that either the principal $\mathbf{Z}[\zeta_p]$-ideal generated by $H(\zeta_p) + q$ or the one generated by $H(\zeta_p) + q^2$ is equal to the prime ideal $\mathfrak{q}$ times a product of prime ideals of *strictly smaller* degree. Therefore, the ideal class of $\mathfrak{q}$ is contained in the subgroup of the class group that is generated by prime ideals of degree $< f$.

It follows by induction that $\mathfrak{q}$ is in the subgroup of the class group that is generated by prime ideals of degree 1. Since $\mathfrak{q}$ is an arbitrary prime ideal of degree ≥ 2, we conclude that the class group is generated by prime ideals of degree 1. Theorem 9.5 implies now that the Stickelberger ideal annihilates the ideal class group, as required.

Exercises

9.1 Let G be a finite abelian group and $G^\vee$ denote its group of characters χ: $G \longrightarrow \mathbf{C}^*$.

(a) Show that the map $C[G] \longrightarrow \prod_{\chi \in G^\vee} C$ defined by $\theta \mapsto (\chi(\theta))_{\chi \in G^\vee}$ is an isomorphism of rings.

(b) Show that any ideal I of $\prod_{\chi \in G^\vee} C$ has the form $\prod_{\chi \in A} C \times \prod_{\chi \notin A} \{0\}$ for some subset A of $G^\vee$.

(c) Let $\theta \in C[G]$. Show that the ideal generated by θ has dimension $m = \#\{\chi \in G^\vee : \chi(\theta) \neq 0\}$.

9.2 Let p be an odd prime and $G = \mathrm{Gal}(Q(\zeta_p)/Q)$. Show that in the ring $Z[G]$ we have $\theta_0 = \theta_1 = 0$ and $\theta_{-1} = -T$.

9.3 For $x, y \in R$, show that $[x] + [y] \leq [x + y] < [x] + [y] + 1$. Let p be an odd prime and $G = \mathrm{Gal}(Q(\zeta_p)/Q)$. Show that the coefficients $m_{i,a}$ of the elements $f_i \in Z[G]$ are either 0 or 1.

9.4 For $x \in R$, show that $[x] + [-x] = 0$ when $x \in Z$, while $[x] + [-x] = -1$ otherwise. Let p be an odd prime and let $f_i = \theta_{i+1} - \theta_i$ be the elements in the group ring $Z[G]$ defined in this chapter. Show that for every $0 \leq i \leq (p-1)/2$, the element $(1 + \iota) f_i$ is equal to the G-trace T.

9.5 Let $p = 11$ and $G = (Z/pZ)^*$. Compute the elements $\theta_i \in Z[G]$ for $1 \leq i \leq p - 1$. Compute the elements $f_i, e_i \in Z[G]$ for $1 \leq i \leq (p-1)/2$.

9.6 Let l and p be odd primes. Let $\tau(\chi)$ be the Gaussian sum associated with the character $\chi : (Z/lZ)^* \longrightarrow \mu_p$. Using the notation of this chapter, show that

(a) the product of $\tau(\chi)$ by its complex conjugate is equal to l,

(b) $\sigma_a(\tau(\chi)) = \tau(\chi^a)$ for all $a \in (Z/pZ)^*$,

(c) $\rho_b(\tau(\chi)) = \chi^{-1}(b)\tau(\chi)$ for all $b \in (Z/lZ)^*$.

9.7 *(Kummer's lemma)*. Let $f \in Z[X]$ be an irreducible monic polynomial. Let α denote a zero of f. Suppose that the ring $Z[\alpha]$ is the ring of integers of the number field $F = Q(\alpha)$. For every prime number l, let $\prod_{g|f \pmod l} g^{e(g)}$ be the factorization of $f \pmod l$ into a product of powers of distinct, irreducible polynomials g. Show that

$$l = \prod_{g|f \pmod l} \mathfrak{l}^{e(g)},$$

where $\mathfrak{l}$ is the $Z[\alpha]$-ideal generated by l and $g(\alpha)$.

9.8 Let $n \in Z_{\geq} 1$. Let F be a number field and $a \in F^*$. Show that any ideal $\mathfrak{p}$ of F that is coprime to n is *unramified* in the extension $F \subset F(\sqrt[n]{a})$ if and only if $\mathrm{ord}_\mathfrak{p}(a) \equiv 0 \pmod n$.

9.9 *(Kummer theory)* Let K be a field and let n be a positive integer not divisible by the characteristic of K. Set $E = K(\zeta_n)$ and $G = \mathrm{Gal}(E/K)$. Let A be a finitely generated subgroup of E^* with the property that the image $A/(A \cap E^{*n})$ is a $Z[G]$-submodule of E^*/E^{*n}. By $E(\sqrt[n]{A})$ we denote the extension of E generated by the nth roots of the elements of A.

(a) Show that $E(\sqrt[n]{A})$ is a finite Galois extension of K containing $E = K(\zeta_n)$ as a subfield.

(b) Consider the Kummer map $E^*/E^{*n} \longrightarrow \mathrm{Hom}(G_E, \mu_n)$ given by $a \mapsto f$, where $f(\sigma) = \sigma(\sqrt[n]{a})/\sqrt[n]{a}$ for all $\sigma \in G_E$. Show that it is an isomorphism of $\mathbf{Z}[G]$-modules. Here $G_E = \mathrm{Gal}(\overline{E}/E)$ and the G-action on $\mathrm{Hom}(G_E, \mu_n)$ is defined by $(\tau f)(\sigma) = \sigma(f(\tau^{-1}\sigma\tau))$ for any $\sigma \in G_E$.

(c) Show that the isormorphism of part (a) induces a G-isomorphism

$$A/(A \cap E^{*n}) \longrightarrow \mathrm{Hom}(\mathrm{Gal}(E(\sqrt[n]{A})/E), \mu_n).$$

(d) Consider the "dual" Kummer isomorphism

$$\mathrm{Gal}(E(\sqrt[n]{A})/E) \longrightarrow \mathrm{Hom}(A/(A \cap E^{*n}), \mu_n)$$

given by $\sigma \mapsto \varphi$, where $\varphi(a) = \sigma(\sqrt[n]{a})/\sqrt[n]{a}$ for all $a \in A$. Show that this is an isomorphism of $\mathbf{Z}[G]$-modules.

(e) Show that every finite extension $E \subset L$ that is Galois over K and for which $\mathrm{Gal}(L/E)$ has an exponent dividing n is of the form $L = E(\sqrt[n]{A})$ for some finitely generated subgroup $A \subset E^*$ with the property that $A/(A \cap E^{*n})$ is a $\mathbf{Z}[G]$-submodule of E^*/E^{*n}.

(f) Suppose that $A \subset K^*$. Show that $\tau\sigma\tau^{-1} = \sigma^{\omega(\tau)}$ for all σ in the Galois group $\mathrm{Gal}(E(\sqrt[n]{A})/E)$ and all $\tau \in G$. Here $\omega : G \longrightarrow (\mathbf{Z}/n\mathbf{Z})^*$ denotes the character given by $\tau(\zeta) = \zeta^{\omega(\tau)}$ for all $\tau \in G$.

(g) Show that if the extension $K \subset E(\sqrt[n]{A})$ is abelian, we have $\tau(a) \equiv a^{\omega(\sigma)} \pmod{E^{*n}}$ for all $a \in A$ and all $\tau \in G$.

10
The Double Wieferich Criterion

When $x, y \in \mathbf{Z}$ are nonzero solutions to Catalan's equation $x^p - y^q = 1$, Cassels' theorem (Theorem 6.4) says that q divides x and p divides y. In this chapter, we prove a stronger version of this statement: We show that q^2 divides x and p^2 divides y. Using this result, one easily proves Theorem I of chapter 1.

Let p be a prime and $G = \text{Gal}(\mathbf{Q}(\zeta_p)/\mathbf{Q})$. The *Stickelberger ideal* of the group ring $\mathbf{Z}[G]$ was introduced in chapter 9. The $\mathbf{Z}[G]$-ideal I was also introduced there. It is defined as the product of the Stickelberger ideal and the element $1 - \iota$. Here $\iota \in G$ denotes complex conjugation, i.e., we have $\iota(\zeta_p) = \zeta_p^{-1}$. We recall the definition of the obstruction group H. We have

$$H = \{\alpha \in \mathbf{Q}(\zeta_p)^* : \text{ord}_{\mathfrak{r}}(\alpha) \equiv 0 \ (\text{mod } q) \text{ for all primes } \mathfrak{r} \neq \mathfrak{p}\}/\mathbf{Q}(\zeta_p)^{*q}.$$

Here $\mathfrak{p}$ denotes the prime ideal $(1 - \zeta_p)$ of the ring $\mathbf{Z}[\zeta_p]$. See chapter 7 for the basic properties of the group H.

Proposition 10.1 *Let p, q be two distinct odd primes. Then the $\mathbf{Z}[G]$-ideal I defined above annihilates the obstruction group H defined in chapter 7.*

Proof Recall the exact sequence

$$0 \longrightarrow E_p/E_p^q \longrightarrow H \longrightarrow Cl_p[q] \longrightarrow 0$$

of chapter 7. Theorem 9.6 says that the Stickelberger ideal *annihilates* the class group and hence the group $Cl_p[q]$. By Lemma 7.1 *(iii)*, the element $1 - \iota$ annihilates E_p/E_p^q. It follows that the ideal I annihilates H, as required.

Theorem 10.2 *Suppose that p, q are odd primes and that x, y are nonzero integers satisfying Catalan's equation $x^p - y^q = 1$. Then q^2 divides x and p^2 divides y.*

Proof By Lemma 6.1, we have $p \neq q$. By symmetry, it suffices to show that q^2 divides x. Let θ be an element of the ideal I introduced above. By Proposition 7.2, the number $x - \zeta_p$ is contained in the obstruction group H. Since I annihilates H, we

R. Schoof, *Catalan's Conjecture*, DOI: 10.1007/978-1-84800-185-5_10,
© Springer-Verlag London Limited 2008

have $(x - \zeta_p)^\theta = \alpha^q$ for some $\alpha \in \mathbf{Q}(\zeta_p)^*$. Multiplying by the qth power $(-\zeta_p^{-1})^\theta$ and applying ι, we find that $(1 - \zeta_p x)^\theta = \beta^q$ for some $\beta \in \mathbf{Q}(\zeta_p)^*$. By Cassels' theorem, q divides x. This shows that both elements $(1 - \zeta_p x)^\theta$ and β are integral at the primes of $\mathbf{Z}[\zeta_p]$ that lie over q. Therefore, it makes sense to speak of their images in the quotient ring $\mathbf{Z}[\zeta_p]/(q)$ or $\mathbf{Z}[\zeta_p]/(q^2)$. The congruences below are to be understood in this sense. Since $x \equiv 0 \pmod{q}$, we have $1 \equiv \beta^q \pmod{q}$. Both sides are qth powers in $\mathbf{Q}(\zeta_p)$. Therefore, Exercises 10.1 and 10.2 imply that the congruence holds modulo q^2. It follows that

$$(1 - \zeta_p x)^\theta \equiv 1 \pmod{q^2}.$$

We write $\theta = \sum_{\sigma \in G} n_\sigma \sigma$ and expand the product $(1 - \zeta_p x)^\theta = \prod_{\sigma \in G}(1 - \sigma(\zeta_p)x)^{n_\sigma}$. Since $x \equiv 0 \pmod{q}$, it follows that

$$1 - \sum_{\sigma \in G} n_\sigma \sigma(\zeta_p)x \equiv 1 \pmod{q^2}.$$

If q^2 does not divide x, we must have $\sum_{\sigma \in G} n_\sigma \sigma(\zeta_p) \equiv 0 \pmod{q}$. Since the roots of unity $\{\sigma(\zeta_p) : \sigma \in G\}$ form a $\mathbf{Z}$-basis for $\mathbf{Z}[\zeta_p]$, this implies that all coefficients n_σ of θ are divisible by q.

Since θ was arbitrary, it follows that every element of the ideal $I \subset \mathbf{Z}[G]$ has all its coefficients divisible by q. But this is not true. Indeed, by Proposition 9.4, the elements $e_i \in I$ have the property that *all* their coefficients are equal to ± 1. For instance, the element $e_2 = (1 - \iota)f_2 = (1 - \iota)\theta_2 \in I$ is equal to

$$- \sum_{1 \le a \le \frac{p-1}{2}} \sigma_a^{-1} + \sum_{\frac{p+1}{2} \le a \le p-1} \sigma_a^{-1}.$$

Therefore, q^2 must divide x and the theorem follows.

Theorem I follows from this.

Theorem I *(P. Mihăilescu, 2000) Suppose that p, q are odd primes and that x, y are nonzero integers satisfying Catalan's equation $x^p - y^q = 1$. Then we have*

$$p^{q-1} \equiv 1 \pmod{q^2} \quad \text{and} \quad q^{p-1} \equiv 1 \pmod{q^2}.$$

Proof By Corollary 6.5 of Cassels' theorem, we have $x - 1 = p^{q-1}a^q$ for a certain $a \in \mathbf{Z}$. Since $x \equiv 0 \pmod{q}$ and since $p^{q-1} \equiv 1 \pmod{q}$ by Fermat's little theorem, we have $-1 \equiv a^q \pmod{q}$. Since both -1 and a^q are qth powers in $\mathbf{Z}$, Exercise 10.1 implies that

$$-1 \equiv a^q \pmod{q^2}.$$

Since q^2 divides x, it follows from Theorem 10.2 that

$$-1 \equiv x - 1 = p^{q-1}a^q \equiv p^{q-1}(-1) \pmod{q^2}.$$

This proves the first congruence. The second follows by symmetry.

We conclude this section by phrasing Theorem 10.2 as a statement about the element $x - \zeta_p$ in the group H. Consider the following subgroup of H:

$$S = \{\alpha \in H : \alpha \text{ is a } q\text{-adic } q\text{th power}\}.$$

Here we call $\alpha \in \mathbf{Q}(\zeta_p)^*$ "a q-adic qth power" if it is a qth power in the completion F_q of $\mathbf{Q}(\zeta_p)$ at each of the primes q lying over q. The group S is called a *Selmer group.*[*] We have

$$S = \ker \left(H \longrightarrow \prod_{q|q} F_q^* / F_q^{*q} \right).$$

Corollary 10.3 Suppose that p, q are odd primes and that x, y are nonzero integers satisfying Catalan's equation $x^p - y^q = 1$. Then the image of $x - \zeta_p$ in H is contained in the Selmer group S.

Proof By Theorem 10.2, the element $1 - \zeta_p^{-1} x$ is congruent to 1 modulo q^2. Since q is odd, Exercise 10.4 implies that it is a qth power in the completions of $\mathbf{Q}(\zeta_p)$ at all primes over q. Since $-\zeta_p$ is a qth power in $\mathbf{Q}(\zeta_p)$, we deduce that $x - \zeta_p$ is "a q-adic qth power", as required.

Exercises

10.1 Let R be a commutative ring and let q be a prime number with the property that R/qR has no nonzero nilpotent elements. Show that if $a, b \in R$ satisfy $a^q \equiv b^q \pmod{qR}$, then they also satisfy $a^q \equiv b^q \pmod{q^2 R}$.

10.2 Let n be a natural number and q be an odd prime. Show that the ring $\mathbf{Z}[\zeta_n]/(q)$ contains nonzero nilpotent elements if and only if q divides n.

---

[*]The group H is equal to the *flat* cohomology group $H^1_{\text{flat}}(X, \mu_q)$. Here μ_q is the group scheme of the qth roots of unity and X denotes the scheme $\text{Spec}(\mathbf{Z}[\zeta_p, \frac{1}{p}])$. The subgroup of $\alpha \in H$ with the property that the extension $\mathbf{Q}(\zeta_p) \subset \mathbf{Q}(\zeta_p, \sqrt[q]{\alpha})$ is unramified at the primes q lying over q makes up the *étale* cohomology group $H^1_{\text{et}}(X, \mu_q)$. Since the elements α contained in the subgroup S have this property, the group S can be identified with a subgroup of $H^1_{\text{et}}(X, \mu_q)$. More precisely, it is the Selmer group

$$\ker \left(H^1_{\text{et}}(X, \mu_q) \longrightarrow \prod_{q|q} H^1(\text{Spec}(F_q), \mu_q) \right).$$

10.3 Let q be an odd prime and O be a finite unramified extension of the ring $\mathbf{Z}_q$ of p-adic integers. Show that the multiplicative group $(1 + qO)^q$ is equal to $1 + q^2O$.

10.4 Let p, q be odd primes and suppose that x, y are nonzero integers that satisfy Catalan's equation $x^p - y^q = 1$. Deduce from Corollary 6.5 that the following three statements are equivalent:

(a) $x - \zeta_p$ is contained in the subgroup S of H defined above;

(b) q^2 divides x;

(c) $p^{q-1} \equiv 1 \pmod{q^2}$.

11
The Minus Argument

In this chapter, we prove Theorem III of chapter 1. The proof is "Archimedean" in the sense that it exploits the fact that for any nonzero solution x,y of Catalan's equation $x^p - y^q = 1$, the absolute values of x and y are necessarily very large. This follows from Corollary 6.5 (iii). We use the notation introduced in chapter 7.

Lemma 11.1 *Let p,q be distinct odd primes and $G = \mathrm{Gal}(\mathbf{Q}(\zeta_p)/\mathbf{Q})$. Suppose that x,y are nonzero integers satisfying Catalan's equation $x^p - y^q = 1$. Then the map*

$$\{\theta \in \mathbf{Z}[G] : (x - \zeta_p)^\theta \in \mathbf{Q}(\zeta_p)^{*q}\} \quad \longrightarrow \quad \mathbf{Q}(\zeta_p)^*$$

that maps θ to the element $\alpha \in \mathbf{Q}(\zeta_p)^*$ for which

$$(x - \zeta_p)^\theta = \alpha^q$$

is an injective homomorphism.

Proof Since $\mathbf{Q}(\zeta_p)$ does not contain any primitive qth roots of unity, the element α is unique and the map is a well defined homomorphism. Suppose $\theta = \sum_{\sigma \in G} n_\sigma \sigma$ is in the kernel. This means that $\prod_{\sigma \in G}(x - \sigma(\zeta_p))^{n_\sigma} = 1$. The ideals $(x - \sigma(\zeta_p))$ have at most the prime factor $1 - \zeta_p$ in common. In fact, by Corollary 6.5, we have $x \equiv 1 \pmod{p}$. Therefore, the ideals $(x - \sigma(\zeta_p))$ actually do have this factor in common. It follows that the principal ideals generated by

$$\frac{x - \sigma(\zeta_p)}{1 - \zeta_p}, \qquad (\sigma \in G),$$

are mutually coprime. We estimate their norms from the field $\mathbf{Q}(\zeta_p)$ to $\mathbf{Q}$. By Corollary 6.5 (iii), we have $|x| \geq q^{p-1} + q$, so that $|x - \sigma(\zeta_p)| > q^{p-1}$ for each $\sigma \in G$. It follows that for all $\sigma \in G$, we have

$$N\left(\frac{x - \sigma(\zeta_p)}{1 - \zeta_p}\right) \geq \frac{q^{(p-1)^2}}{p} > 1.$$

R. Schoof, *Catalan's Conjecture*, DOI: 10.1007/978-1-84800-185-5_11,
© Springer-Verlag London Limited 2008

This shows that for every $\sigma \in G$, the ideal $(x - \sigma(\zeta_p))$ is divisible by a prime ideal that does not divide any of the other factors in the product $\prod_{\sigma \in G}(x - \sigma(\zeta_p))^{n_\sigma} = 1$. It follows that n_σ is zero for all σ.

We conclude that the homomorphism is injective, as required.

Recall that by Proposition 7.2, the element $x - \zeta_p$ modulo $\mathbf{Q}(\zeta_p)^{*q}$ is contained in the obstruction group

$$H = \left\{ \alpha \in \mathbf{Q}(\zeta_p)^* : \mathrm{ord}_\tau(\alpha) \equiv 0 \ (\mathrm{mod}\ q) \text{ for all primes } \tau \neq \mathfrak{p} \right\} / \mathbf{Q}(\zeta_p)^{*q}$$

that was defined in chapter 7. The left-hand group in Lemma 11.1 is precisely the $\mathbf{Z}[G]$-annihilator of the element $x - \zeta_p$ in H.

Definition For any element $\theta = \sum_{\sigma \in G} n_\sigma \sigma$ of the group ring $\mathbf{Z}[G]$, we define its *size* $\|\theta\|$ by

$$\|\theta\| = \sum_{\sigma \in G} |n_\sigma|.$$

Proposition 11.2 *Let p,q be distinct odd primes and let G be the Galois group of $\mathbf{Q}(\zeta_p)$ over $\mathbf{Q}$. Suppose that $x \in \mathbf{Z}$ satisfies $|x| \geq 2$ and that $(x - \zeta_p)^\theta = \alpha^q$ for some θ in the ideal $(1 - \iota) \subset \mathbf{Z}[G]$ and some $\alpha \in \mathbf{Q}(\zeta_p)^*$. Then for every embedding $\phi : \mathbf{Q}(\zeta_p) \hookrightarrow \mathbf{C}$, the number $\phi(\alpha)$ is contained in the unit circle and there exists a qth root of unity $\xi \in \mathbf{C}$ for which*

$$|\phi(\alpha) - \xi| \leq \frac{\|\theta\|}{q(|x| - 1)}.$$

Proof Let $\phi : \mathbf{Q}(\zeta_p) \hookrightarrow \mathbf{C}$ be a fixed embedding. Since $\theta = \sum_{\sigma \in G} n_\sigma \sigma$ is in the ideal $(1 - \iota)$ of $\mathbf{Z}[G]$, the absolute value of $\phi((x - \zeta_p)^\theta)$ is equal to 1. Therefore, the same is true for $\phi(\alpha)$, so that it lies on the unit circle.

Using the Taylor series expansion of the principal branch of the logarithm, Exercise 11.1 implies

$$|\mathrm{Arg}(\phi(\alpha)^q)| = |\log(\phi(\alpha)^q)| \leq \sum_{\sigma \in G} |n_\sigma| \left|\log\left(1 - \frac{\phi(\sigma(\zeta))}{x}\right)\right| \leq \frac{\|\theta\|}{|x| - 1},$$

and hence

$$\left|\mathrm{Arg}(\phi(\alpha)) - \frac{2\pi k}{q}\right| \leq \frac{\|\theta\|}{q(|x| - 1)} \qquad \text{for some } k \in \mathbf{Z}.$$

The left-hand side of this inequality is precisely the length of the arc between the complex numbers $\phi(\alpha)$ and $\xi = e^{\frac{2\pi i k}{q}}$. Since the *arc* is at least as long as the *segment* between these two numbers, the proposition easily follows.

In chapter 10, we defined the ideal $I \subset \mathbf{Z}[G]$ as the product of the Stickelberger ideal by $1 - \iota$. If $x, y \in \mathbf{Z}$ is a nonzero solution of Catalan's equation, Proposition 10.1 implies that every $\theta \in I$ satisfies the condition $(x - \zeta_p)^\theta = \alpha^q$ of Proposition 11.2. This explains the importance of the ideal I for the proof of Theorem III. Under the assumption $q > 4p^2$, Mihăilescu shows that the ideal $I \subset \mathbf{Z}[G]$ contains *many* elements θ of small size. This is the content of the next proposition. We will then use the box principle to derive a contradiction from this and deduce Theorem III.

Proposition 11.3 *Let $p, q \geq 5$ be prime numbers satisfying $q > 4p^2$. Then the ideal I contains at least $q + 1$ elements of size at most $\frac{3}{2}\frac{q}{p-1}$.*

Proof Let $s = [\frac{3}{2}\frac{q}{(p-1)^2}]$. By Proposition 9.3 *(ii)*, the elements $e_i = (1-\iota)f_i$ for $i = 1, 2, \ldots, \frac{p-1}{2}$ form a $\mathbf{Z}$-basis for the ideal I introduced above. By Proposition 9.4, half of the coefficients of e_i are equal to 1, while the other ones are all equal to -1. Therefore, the size $\|e_i\|$ of e_i is equal to $p - 1$ for each $1 \leq i \leq (p - 1)/2$. It follows that any element of the form

$$\theta = \sum_{i=1}^{(p-1)/2} \lambda_i e_i,$$

with $\lambda_i \in \mathbf{Z}_{\geq 0}$ satisfying $\sum_{i=1}^{(p-1)/2} \lambda_i \leq s$, has the property that

$$\|\theta\| \leq (p - 1) \sum_{i=1}^{\frac{p-1}{2}} \lambda_i \leq \frac{3}{2}\frac{q}{p - 1}.$$

By Exercise 11.3, there are $\binom{s+\frac{p-1}{2}}{s}$ such elements. There are exactly as many elements $\sum_{i=1}^{(p-1)/2} \lambda_i e_i$ with all coefficients $\lambda_i \leq 0$ that satisfy $\sum_{i=1}^{(p-1)/2} |\lambda_i| \leq s$. Therefore, there are at least

$$2\binom{s + \frac{p-1}{2}}{s} - 1$$

elements in I of size at most $\frac{3}{2}\frac{q}{p-1}$.

Now we apply Lemma 11.4 below with $k = \frac{p-1}{2}$. Since $s = [\frac{3}{2}\frac{q}{(p-1)^2}]$, this leads to the inequality

$$2\binom{s + \frac{p-1}{2}}{s} - 1 \geq \frac{2}{3}(s + 1)(p - 1)^2 + 1 \geq q + 1.$$

This shows that there are at least $q + 1$ elements θ, as required.

It remains to check that the conditions of Lemma 11.4 are satisfied: Since $q > 4p^2$, the number $s = [\frac{3}{2}\frac{q}{(p-1)^2}]$ is at least 6. Since $p \geq 5$, we have $k \geq 2$.

Moreover, (k, s) is not equal to any of the exceptional pairs that are listed in Lemma 11.4. Indeed, if $k = 2$, we have $p = 5$ and $100 = 4p^2 < q \leq \frac{2}{3}(s + 1)$ $(p-1)^2 = 32(s+1)/3$. This implies, $s > 300/32 > 9$. If $k = 3$, we have $p = 7$ and $196 = 4p^2 < q \leq \frac{2}{3}(s + 1)(p - 1)^2 = 24(s + 1)$, which implies $s > 196/24 > 8$.

   This completes the proof of the proposition.

Lemma 11.4 *For all pairs of integers (k, s) with $k \geq 2$ and $s \geq 6$ with the exception of $(k, s) = (2, 6), (2, 7), (2, 8)$, and $(3, 6)$, we have*

$$\binom{s + k}{s} > \frac{4}{3}(s + 1)k^2 + 1.$$

Proof See Exercise 11.4.

Corollary 11.5 *Let p, q be odd primes satisfying $q > 4p^2$ and suppose that $x, y \in$ $\mathbf{Z}$ is a nonzero solution of Catalan's equation $x^p - y^q = 1$. Then for any embedding $\phi : \mathbf{Q}(\zeta_p) \hookrightarrow \mathbf{C}$, there exists a nonzero element $\theta \in I$ with $\|\theta\| \leq \frac{3q}{p-1}$ with the property that $(x - \zeta_p)^\theta = \alpha^q$ for some $\alpha \in \mathbf{Q}(\zeta_p)$ with*

$$|\phi(\alpha) - 1| \leq \frac{2\|\theta\|}{q(|x| - 1)}.$$

Proof By Proposition 7.2, the number $x - \zeta_p$ modulo $\mathbf{Q}(\zeta_p)^{*q}$ is contained in the obstruction group H. By Proposition 10.1, the ideal I annihilates H. Proposition 11.3 ensures the existence of at least $q + 1$ elements θ in I of size at most $\frac{3}{2}\frac{q}{p-1}$. For each of these elements $\theta \in I$, we have $(x - \zeta_p)^\theta = \alpha^q$ for some $\alpha \in \mathbf{Q}(\zeta_p)$ depending on θ; hence, by Proposition 11.2,

$$|\phi(\alpha) - \xi| \leq \frac{\|\theta\|}{q(|x| - 1)},$$

for some qth root of unity $\xi \in \mathbf{C}$ that depends on θ. By the box principle, there exist then two *distinct* such elements θ_1, θ_2 satisfying this inequality for α_1, α_2 with *the same* qth root of unity $\xi = \xi_1 = \xi_2$. We set $\theta = \theta_1 - \theta_2$ and $\alpha = \alpha_1/\alpha_2$. Then we have $(x - \zeta_p)^\theta = \alpha^q$. Moreover, we have $\theta \neq 0$ and, by Exercise 11.2, we have $\|\theta\| \leq \|\theta_1\| + \|\theta_2\| \leq \frac{3q}{p-1}$. We also have

$$
\begin{aligned}
|\phi(\alpha) - 1| &= |\phi(\alpha_1) - \phi(\alpha_2)| \\
&\leq |\phi(\alpha_1) - \xi| + |\phi(\alpha_2) - \xi| \\
&\leq \frac{2\|\theta\|}{q(|x| - 1)},
\end{aligned}
$$

as required.

Theorem III *(P. Mihăilescu, 2003) For any pair of odd primes p,q with q > $4p^2$ or p > $4q^2$, there are no nonzero x, y ∈ **Z** for which Catalan's equation $x^p - y^q = 1$ holds.*

Proof By Theorem IV, we may assume $p, q \geq 5$. The proof is by contradiction. Suppose that x,y are nonzero integers that satisfy Catalan's equation $x^p - y^q = 1$. By symmetry, we may assume that $q > 4p^2$. We fix an embedding $\phi : \mathbf{Q}(\zeta_p) \hookrightarrow \mathbf{C}$. By Corollary 11.5, there exists a nonzero element $\theta \in I$ with $\|\theta\| \leq \frac{3q}{p-1}$ with the properties

$$(x - \zeta_p)^\theta = \alpha^q \qquad \text{for some } \alpha \in \mathbf{Q}(\zeta_p)$$

and

$$|\phi(\alpha) - 1| \quad \leq \quad \frac{2\|\theta\|}{q(|x| - 1)}.$$

The same inequality holds for the complex conjugate of $\phi(\alpha)$. For the remaining $p - 3$ embeddings $\tilde{\phi}$, we have $|\tilde{\phi}(\alpha) - 1| \leq 2$. This follows from the fact that $\tilde{\phi}(\alpha) = 1$ for every embedding $\tilde{\phi}$. Taking the product over the embeddings $\phi : \mathbf{Q}(\zeta_p) \hookrightarrow \mathbf{C}$, we obtain the following inequality for the norm from $\mathbf{Q}(\zeta_p)$ to $\mathbf{Q}$ of $\alpha - 1$:

$$N(\alpha - 1) \quad \leq \quad \frac{2^{p-1}}{q^2} \left(\frac{\|\theta\|}{|x| - 1} \right)^2.$$

In order to obtain a *lower bound* for $N(\alpha - 1)$, we first note that $\alpha - 1$ is not zero. This follows from Lemma 11.1 and the fact that $\theta \neq 0$. We estimate the norm of its denominator J. The ideal J of $\mathbf{Z}[\zeta_p]$ is also the denominator of α. So, the principal fractional ideal (α) is equal to J'/J for some $\mathbf{Z}[\zeta_p]$-ideal J' that is coprime to J. It is convenient to take qth powers and to consider the fractional ideal $(\alpha)^q = (x - \zeta_p)^\theta$. We have

$$(x - \zeta_p)^\theta = \prod_{\sigma \in G}(x - \sigma(\zeta_p))^{n_\sigma} = J'^q/J^q.$$

Since $\theta = (1 - \iota)\psi$ for some ψ, the norm of $(x - \zeta_p)^\theta$ is equal to 1. It follows that $N(J') = N(J)$. We also have

$$\prod_{\sigma \in G}(x - \sigma(\zeta_p))^{|n_\sigma|} = J'^q J^q.$$

This implies that $N(J)^{2q} = N(J'J)^q$ divides $N(\prod_{\sigma \in G}(x - \zeta_p)^{|n_\sigma|})$, which is at most $(|x| + 1)^{(p-1)\|\theta\|}$. Therefore,

$$N(J) \quad \leq \quad (|x| + 1)^{\frac{p-1}{2q}\|\theta\|}.$$

On the other hand, since $J(\alpha - 1)$ is an *integral* nonzero ideal, we have $N(J(\alpha - 1)) \geq 1$. Combining everything, we obtain the inequality

$$(|x| + 1)^{-\frac{p-1}{2q}\|\theta\|} \quad \leq \quad N(J)^{-1} \quad \leq \quad N(\alpha - 1) \quad \leq \quad \frac{2^{p-1}}{q^2}\left(\frac{\|\theta\|}{|x| - 1}\right)^2.$$

Since $|x| \geq q^{p-1} > 20$, we have $|x| + 1 \leq \frac{4}{3}(|x| - 1)$ and hence

$$(|x| + 1)^{2 - \frac{p-1}{2q}\|\theta\|} \quad \leq \quad \frac{16}{9}\frac{\|\theta\|^2}{q^2}2^{p-1}. \qquad\qquad (*)$$

The size of θ satisfies $\|\theta\| \leq \frac{3q}{p-1}$, so that

$$(|x| + 1)^{1/2} \quad \leq \quad \frac{16}{(p-1)^2}2^{p-1}.$$

By Corollary 6.5, we have $|x| > q^{p-1}$. Since $p \geq 5$, this leads to

$$q^{(p-1)/2} \quad < \quad (|x| + 1)^{1/2} \quad \leq \quad 2^{p-1},$$

contradicting $q \geq 5$.

This proves the theorem.

In the proof of Theorem III, it is essential that the exponent $2 - \frac{p-1}{2q}\|\theta\|$ to which $|x| + 1$ is raised in the left-hand side of the inequality $(*)$ is *positive*. Here the positive contribution "2" in the exponent is the number of embeddings $\mathbf{Q}(\zeta_p) \hookrightarrow \mathbf{C}$ for which the absolute value of the image of $\alpha - 1$ is *very small*. To guarantee this, we made the assumption $\|\theta\| \leq \frac{3q}{p-1}$, but the choice of the constant 3 is rather arbitrary. With some more work, one can prove a similar theorem for a somewhat larger constant.

Exercises

11.1 Let "log" denotes the principal branch of the complex logarithm function.

  (a) Show that $|\log(1 + z)| < \frac{|z|}{1-|z|}$ for all $z \in \mathbf{C}$ with $|z| < 1$.

  (b) Show that $|\log(xy)| \leq |\log(x)| + |\log(y)|$ for all complex numbers x, y that are not contained in $\mathbf{R}_{<0}$.

11.2 Let p be a prime and let $G = \mathrm{Gal}(\mathbf{Q}(\zeta_p)/\mathbf{Q})$.

  (a) For $\theta \in \mathbf{Z}[G]$, show that $\|\theta\| = 0$ if and only if $\theta = 0$.

  (b) Show that $\|\theta + \theta'\| \leq \|\theta\| + \|\theta'\|$ for all $\theta, \theta' \in \mathbf{Z}[G]$.

11.3 Let k and s be natural numbers. Show that the number of row vectors $(\lambda_1, \ldots, \lambda_k)$ with $\lambda_i \in \mathbf{Z}$ satisfying $0 \leq \lambda_i \leq s$ and $\sum_{j=1}^{k} \lambda_j \leq s$ is equal to the binomial coefficient $\binom{s+k}{s}$.

11.4 The goal of this exercise is to prove the inequality of Lemma 11.4.

    (a) Show that the inequality

$$\binom{s+k}{s} > \frac{4}{3}(s+1)k^2 + 1$$

    holds for the pairs $(k, s) = (4, 6)$, $(3, 7)$, and $(2, 9)$.

    (b) Show inductively that if the inequality holds for a pair (s, k), it also holds for all pairs s', k' for which $s' \geq s$ and $k' \geq k$.

    (c) Prove Lemma 11.4.

12
The Plus Argument I

In this chapter and chapter 14, we prove Theorem II of chapter 1. There are two main ingredients: the Runge method, exploited here, and Thaine's theorem, which is used in chapter 14. While the proofs of the results of chapters 8, 10, and 11 involve objects like the minus class group and the Stickelberger ideal, both *anti-invariant* under complex conjugation, the present proof involves *invariant* objects like the group of cyclotomic units and the class group Cl_p^+ of the field $\mathbf{Q}(\zeta_p^+)$.

Let p, q be distinct odd primes. We use the notation of Chapter 7. In particular, we have $G = \mathrm{Gal}(\mathbf{Q}(\zeta_p)/\mathbf{Q})$ and $G^+ = \mathrm{Gal}(\mathbf{Q}(\zeta_p^+)/\mathbf{Q})$. For any $s \in \mathbf{Q}$, we let $(1 + T)^s$ denote the power series in $\mathbf{Q}[[T]]$ defined by

$$(1 + T)^s = \sum_{k \geq 0} \binom{s}{k} T^k.$$

We formulate two propositions concerning power series. For a natural number q and an element $\theta = \sum_{\sigma \in G} n_\sigma \sigma$ in the group ring $\mathbf{Z}[G]$, let

$$F(T) = (1 - \zeta_p T)^{\theta/q} = \prod_{\sigma \in G}(1 - \sigma(\zeta_p)T)^{n_\sigma/q} \in \mathbf{Q}(\zeta_p)[[T]],$$

where, as explained above, each factor in the product is defined by

$$(1 - \sigma(\zeta_p)T)^{n_\sigma/q} = \sum_{k \geq 0} \binom{n_\sigma/q}{k}(-\sigma(\zeta_p)T)^k \in \mathbf{Q}(\zeta_p)[[T]].$$

For each embedding $\phi : \mathbf{Q}(\zeta_p) \hookrightarrow \mathbf{C}$, we let $F^\phi(T)$ denote the power series in $\mathbf{C}[[T]]$ obtained by applying ϕ to the coefficients of $F(T)$. For any $t \in \mathbf{C}$, substituting $T = t$ in the power series $F^\phi(T)$, we obtain a series of complex numbers. If this series converges, we denote its sum by $F^\phi(t)$.

Proposition 12.1 *Let $\theta = \sum_{\sigma \in G} n_\sigma \sigma \in \mathbf{Z}[G]$. Using the notation above, we have the following.*

(i) The coefficients of $F(T)$ are integral outside q.

R. Schoof, *Catalan's Conjecture*, DOI: 10.1007/978-1-84800-185-5_12,
© Springer-Verlag London Limited 2008

(ii) The power series $F(T)$ has the form

$$F(T) = \sum_{k \geq 0} \frac{c_k}{k!q^k} T^k,$$

with $c_k \in \mathbf{Z}[\zeta_p]$ satisfying

$$c_k \equiv (-\sum_{\sigma \in G} \sigma(\zeta_p)n_\sigma)^k \pmod{q}.$$

(iii) For any embedding $\phi : \mathbf{Q}(\zeta_p) \hookrightarrow \mathbf{C}$ and any $t \in \mathbf{C}$ with $|t| < 1$, substituting $T = t$ into $F(T)$ leads to a convergent series. For any $k \geq 0$, we have

$$|F^\phi(t) - F_k^\phi(t)| \leq \left|\binom{-m}{k+1}\right| \frac{|t|^{k+1}}{(1-|t|)^{m+k+1}}.$$

Here $F_k^\phi(T)$ is the polynomial that is the sum of the first k terms of $F^\phi(T)$ and $F_k^\phi(t)$ is its value in the complex number t. By m we denote the rational number $\frac{1}{q}\|\theta\| = \frac{1}{q}\sum_{\sigma \in G} |n_\sigma|$.

Proof Part *(i)* follows at once from Exercise 5.7. To prove *(ii)*, we consider the power series

$$G(T) = F(qT) = (1 - \sigma(\zeta_p)qT)^{\frac{n_\sigma}{q}}$$
$$= \sum_{k \geq 0} \frac{n_\sigma(n_\sigma - q)\cdots(n_\sigma - (k-1)q)}{k!}(-\sigma(\zeta_p)T)^k.$$

The coefficients of this power series are visibly of the form $a_k/k!$, where a_k is an element of $\mathbf{Z}[\zeta_p]$ that is congruent to $(-\sigma(\zeta_p)n_\sigma)^k$ (mod q). An application of Exercise 12.2 to the domain $\mathbf{Z}[\zeta_p]$ shows that $G(T)$ is equal to $\sum_{k \geq 0} \frac{c_k}{k!}T^k$ with

$$c_k \equiv \left(-\sum_{\sigma \in G} \sigma(\zeta_p)n_\sigma\right)^k \pmod{q}.$$

Since we have $G(T) = F(qT)$, the result follows.

To prove *(iii)*, we observe that by Exercise 12.1, we have for every $\sigma \in G$, that

$$\left|\binom{n_\sigma/q}{k}\right| \leq \left|\binom{-|n_\sigma|/q}{k}\right| = (-1)^k \binom{-|n_\sigma|/q}{k}.$$

It follows that the absolute values of the coefficients of the series

$$\sum_{k \geq 0} \binom{n_\sigma/q}{k}(-\sigma(\zeta_p)T)^k$$

are less than or equal to the corresponding coefficients of the series

$$\sum_{k\geq 0}\binom{-|n_\sigma|/q}{k}(-T)^k \;\;=\;\; (1-T)^{-\frac{|n_\sigma|}{q}},$$

all of whose coefficients are positive. Therefore, the absolute values of the coefficients of $F^\phi(T)$ are smaller than those of the power series

$$\prod_{\sigma\in G}(1-T)^{-\frac{|n_\sigma|}{q}} = (1-T)^{-\frac{1}{q}\sum_\sigma |n_\sigma|} = (1-T)^{-m} = \sum_{k\geq 0}\binom{-m}{k}(-T)^k.$$

It follows that

$$|F^\phi(t) - F_k^\phi((t)| \;\leq\; |(1-|t|)^{-m} - s_k(|t|)|$$

for $t \in \mathbf{C}$ with $|t| < 1$. Here $s_k(T)$ denotes the sum of the terms of degree at most k of the Taylor series of the function $(1-T)^{-m}$. By Lemma 5.1 and Exercise 5.8, we have

$$|(1-|t|)^{-m} - s_k(|t|)| \leq \left|\binom{-m}{k+1}\right|\frac{|t|^{k+1}}{(1-|t|)^{m+k+1}},$$

which implies the inequality of *(iii)*.

This proves the proposition.

Proposition 12.2 *We use the notation of the previous proposition. If $\theta = \sum_{\sigma\in G} n_\sigma \sigma$ is contained in the ideal $(1 + \iota)$ of $\mathbf{Z}[G]$, then we have the following.*

(i) The power series $F(T)$ is contained in the subring $\mathbf{Q}(\zeta_p^+)[[T]]$ of the power series ring $\mathbf{Q}(\zeta_p)[[T]]$.

(ii) Suppose that $t \in \mathbf{Q}$ satisfies $|t| < 1$ and that $(1 - t\zeta_p)^\theta = \beta^q$ for some element $\beta \in \mathbf{Q}(\zeta_p^+)$. Then we have that $F^\phi(t) = \phi(\beta)$ for all embeddings $\phi : \mathbf{Q}(\zeta_p^+) \hookrightarrow \mathbf{R}$.

Proof The fact that θ is contained in the ideal $(1 + \iota)$ means precisely that for all $\sigma \in G$, the coefficients n_σ and $n_{\sigma\iota}$ of θ are equal. Since, for every $\sigma \in G$, the coefficients of the polynomial $(1 - \sigma(\zeta_p)T)(1 - \sigma\iota(\zeta_p)T)$ are contained in $\mathbf{Q}(\zeta_p^+)$, part *(i)* follows.

We prove *(ii)*. Let ϕ be an embedding $\mathbf{Q}(\zeta_p)^+ \hookrightarrow \mathbf{R}$. We extend it to an embedding $\mathbf{Q}(\zeta_p) \hookrightarrow \mathbf{C}$. Since ϕ is a homomorphism, we have $\phi(\beta)^q = \phi((1 - t\zeta_p)^\theta) = \prod_{\sigma\in G}(1 - t\phi(\sigma(\zeta_p)))^{n_\sigma}$. This number is contained in $\mathbf{R}$. By ordinary analysis, its *real* qth root is equal to $F^\phi(t)$. In other words, we have $\phi(\beta)^q = F^\phi(t)^q$. Since q is odd, real numbers have unique qth roots. Therefore, we have $\phi(\beta) = F^\phi(t)$, as required.

Part *(ii)* of Proposition 12.2 addresses a delicate matter. In general, substitution in power series does not commute with Galois action. But here it does.

In the rest of this chapter, we slightly abuse notation as follows. Consider the natural surjective homomorphism $\mathbf{F}_q[G] \twoheadrightarrow \mathbf{F}_q[G^+]$. For any $\psi \in \mathbf{F}_q[G^+]$, the product of $(1 + \iota)$ by a lift of ψ to $\mathbf{F}_q[G]$ does not depend on the lift of ψ. Therefore, we write $(1 + \iota)\psi$ for this element of $\mathbf{F}_q[G]$.

Before proving the main result of this chapter, we prove a lemma. Recall the definition of the obstruction group H. We have

$$H = \left\{ \alpha \in \mathbf{Q}(\zeta_p)^* : \mathrm{ord}_{\mathfrak{r}}(\alpha) \equiv 0 \ (\mathrm{mod}\ q) \text{ for all primes } \mathfrak{r} \neq \mathfrak{p} \right\} / \mathbf{Q}(\zeta_p)^{*q}.$$

Lemma 12.3 *Let p, q be distinct odd primes and suppose that $x \in \mathbf{Z}$ satisfies $x \equiv 1 \ (\mathrm{mod}\ p)$. Suppose that $\psi \in \mathbf{F}_q[G^+]$ annihilates the element $(x - \zeta_p)^{1+\iota}$ of H^+. Then there exists a lift $\theta = \sum_{\sigma \in G} n_\sigma \sigma \in \mathbf{Z}[G]$ of one of the elements $\pm(1 + \iota)\psi$ with the following properties:*

- $n_\sigma = n_{\sigma\iota}$ *for all $\sigma \in G$;*
- $n_\sigma \geq 0$ *for each $\sigma \in G$;*
- $\sum_\sigma n_\sigma = mq$ *for some integer m satisfying $0 \leq m \leq (p-1)/2$;*
- $(x - \zeta_p)^\theta = \alpha^q$ *for some algebraic integer $\alpha \in \mathbf{Q}(\zeta_p^+)^*.$*

Proof Let $\theta' = \sum_{\sigma \in G} n_\sigma \sigma \in \mathbf{Z}[G]$ be the unique lift of $(1 + \iota)\psi \in \mathbf{F}_q[G]$ with the property that $0 \leq n_\sigma < q$ for every $\sigma \in G$. Notice that automatically $n_\sigma = n_{\sigma\iota}$ for all $\sigma \in G$, so that θ' is contained in the ideal $(1 + \iota)$ of the group ring $\mathbf{Z}[G]$. The element $\theta'' = \sum_{\sigma \in G}(q - n_\sigma)\sigma \in \mathbf{Z}[G]$ lifts $-(1 + \iota)\psi$ and has its coefficients between 1 and q. It is also contained in the ideal $(1 + \iota)$. Both elements θ' and θ'' have nonnegative coefficients. Since we have

$$\theta' + \theta'' = q \sum_{\sigma \in G} \sigma,$$

the sum of the coefficients of $\theta' + \theta'' = q \sum_{\sigma \in G} \sigma$ is equal to $q(p - 1)$. Therefore the sum of the coefficients of one of θ', θ'' is equal to mq for some rational number m satisfying $0 \leq m \leq \frac{p-1}{2}$. That's the element θ we take. By construction, we have

$$(x - \zeta_p)^\theta = \alpha^q \quad \text{for some } \alpha \in \mathbf{Q}(\zeta_p)^*.$$

Since the coefficients n_σ are nonnegative, the number $(x - \zeta_p)^\theta$ is an algebraic integer. Since we have $n_\sigma = n_{\iota\sigma}$ for all $\sigma \in G$, the element $(x - \zeta_p)^\theta$ is invariant under complex multiplication and is hence contained in $\mathbf{Q}(\zeta_p^+)$. The same is then true for α.

Finally, to see that m is an integer, we note that $x \equiv 1 \ (\mathrm{mod}\ p)$, so that $\pi = 1 - \zeta_p$ divides every conjugate of $x - \zeta_p$ exactly once. It follows that

$$\sum_{\sigma \in G} n_\sigma = \mathrm{ord}_\pi((x - \zeta_p)^\theta) = \mathrm{ord}_\pi(\alpha^q) \equiv 0 \ (\mathrm{mod}\ q).$$

This proves the lemma.

To put the next theorem in a context, recall that Theorem 8.3 says that if $x^p - y^q = 1$ is a nontrivial solution of Catalan's equation, then the minus component $(x - \zeta_p)^{1-\iota}$ of the element $x - \zeta_p$ of the obstruction group H is *nontrivial*. We now prove a much stronger statement for the plus component $(x - \zeta_p)^{1+\iota}$: We show that it generates a *free* $\mathbf{F}_q[G^+]$-module.

Theorem 12.4 *Let $p, q \geq 7$ be distinct primes and suppose that $x, y \in \mathbf{Z}$ is a nonzero solution to Catalan's equation. Then the $\mathbf{F}_q[G]$-submodule of H that is generated by the image of $(x - \zeta_p)^{1+\iota}$ is a free module over the ring $\mathbf{F}_q[G^+]$. In other words, the $\mathbf{F}_q[G^+]$-annihilator of $(x - \zeta_p)^{1+\iota}$ is trivial.*

Proof Suppose that $\psi \in \mathbf{F}_q[G^+]$ annihilates the element $(x - \zeta_p)^{1+\iota}$ of H. By Corollary 6.5, we have $x \equiv 1 \ (\mathrm{mod}\ p)$, and Lemma 12.3 applies. Let $\theta = \sum_\sigma n_\sigma \sigma$ in $\mathbf{Z}[G]$ denote the lift of $\pm(1+\iota)\psi$, whose existence is guaranteed by Lemma 12.3. We have therefore

$$(x - \zeta_p)^\theta = \prod_{\sigma \in G}(x - \sigma(\zeta_p))^{n_\sigma} = \alpha^q$$

for some nonzero algebraic integer $\alpha \in \mathbf{Q}(\zeta_p^+)$. Since q is odd and $\mathbf{Q}(\zeta_p^+)$ does not contain any nontrivial qth roots of unity, the element α is unique. We deal with the Diophantine equation

$$\prod_{\sigma \in G}(X - \sigma(\zeta_p))^{n_\sigma} = V^q$$

using Runge's method. By Lemma 12.3, we have that $\sum_{\sigma \in G} n_\sigma = mq$ for some integer $m \geq 0$. It follows that for each embedding $\phi : \mathbf{Q}(\zeta_p) \hookrightarrow \mathbf{C}$, we have

$$\phi(\alpha) = x^m \prod_{\sigma \in G}\left(1 - \frac{\phi(\sigma(\zeta_p))}{x}\right)^{n_\sigma/q} = x^m F^\phi\left(\frac{1}{x}\right).$$

Here $F(T)$ is the power series that occurs in Proposition 12.1 and $F^\phi(T) \in \mathbf{R}[[T]]$ is the power series one obtains by applying ϕ to its coefficients. Let $F_m(T)$ denote the polynomial that is the sum of the terms of $F(T)$ of degree $\leq m$. Consider the element

$$z = \alpha - x^m F_m\left(\frac{1}{x}\right) \in \mathbf{Q}(\zeta_p^+).$$

For every embedding $\phi : \mathbf{Q}(\zeta_p) \hookrightarrow \mathbf{C}$, we have

$$|\phi(z)| = \left|\phi(\alpha) - x^m F_m^\phi\left(\frac{1}{x}\right)\right| = \left|x^m F^\phi\left(\frac{1}{x}\right) - x^m F_m^\phi\left(\frac{1}{x}\right)\right|$$

$$\leq \frac{1}{|x|}\left|\binom{-m}{m+1}\right|\left(1 - \frac{1}{|x|}\right)^{-2m-1} \leq \frac{4^m}{|x|}\left(1 - \frac{1}{|x|}\right)^{-2m-1}.$$

The inequality follows from Proposition 12.1 *(iii)* and the fact that we have $\left|\binom{-m}{m+1}\right| = \binom{2m}{m+1} \leq 4^m$. See Exercise 2.4.

Since the coefficients of the polynomial $F_m(t)$ have denominators, the number $z = \alpha - x^m F_m(\frac{1}{x})$ is not integral. By Proposition 12.1 *(ii)* and *(iii)*, the denominators of the coefficients of $F_m(t)$ divide $D = q^{m+\mathrm{ord}_q(m!)}$. Therefore,

$$Dz = D\left(\alpha - x^m F\left(\frac{1}{x}\right)\right)$$

is integral. By the estimate above, for every embedding $\phi : \mathbf{Q}(\zeta_p^+) \hookrightarrow \mathbf{R}$, we have

$$|\phi(Dz)| \leq q^{m+\mathrm{ord}_q(m!)}\frac{4^m}{|x|}\left(1 - \frac{1}{|x|}\right)^{-2m-1}.$$

Therefore,

$$\log|\phi(Dz)| \leq -\log|x| + \left(m + \frac{m}{q-1}\right)\log q + m\log 4 - (2m+1)\log\left(1 - \frac{1}{|x|}\right).$$

We confront this estimate with the fact that $|x|$ is very large. By Corollary 6.5, we have $|x| \geq q^{p-1} + q$. This implies that

$$-\log|x| \leq -(p-1)\log(q),$$

$$-\log\left(1 - \frac{1}{|x|}\right) \leq \frac{1}{|x|-1} \leq 1/q^{p-1} \leq 1/q^2.$$

See Exercise 11.1. Since we have $0 \leq m \leq (p-1)/2$, this leads to

$$\frac{\log|\phi(Dz)|}{\log(q)} \leq \frac{p-1}{2}\left(-1 + \frac{1}{q-1} + \frac{\log 4}{\log q}\right) + \frac{p}{q^2\log q}.$$

Since $q \geq 7$, this is at most

$$\frac{p-1}{2}\left(-1 + \frac{1}{6} + \frac{\log 4}{\log 7}\right) + \frac{p}{49 \log 7},$$

which is negative for all $p \geq 3$.

This proves that $|\phi(Dz)| < 1$ for every embedding $\phi : \mathbf{Q}(\zeta_p^+) \hookrightarrow \mathbf{R}$. It follows that the absolute value of the norm of Dz is less than 1. Since Dz is an algebraic integer, this implies that Dz is zero. Therefore, we have

$$D\alpha = Dx^m F_m\left(\frac{1}{x}\right)$$

$$= \sum_{k=0}^{m} q^{m+\mathrm{ord}_q(m!)} \frac{c_k}{q^k k!} x^{m-k}.$$

Since α is an algebraic integer, $D\alpha$ is an algebraic integer that is divisible by q. By Proposition 12.1 (ii), all terms on the right-hand side are q-adic integers. Moreover, all are divisible by q, except possibly for the mth one. Then the mth term must, of course, also be divisible by q. This implies that the coefficient c_m is divisible by q. By Proposition. 12.1 (ii), we have then that

$$\left(-\sum_{\sigma \in G} n_\sigma \sigma(\zeta_p)\right)^m \equiv 0 \pmod{q}.$$

By Exercise 10.2, the ring $\mathbf{Z}[\zeta_p]/(q)$ does not contain any nonzero nilpotent elements and it follows that

$$\sum_{\sigma \in G} n_\sigma \sigma(\zeta_p) \equiv 0 \pmod{q}.$$

Since the set $\{\sigma(\zeta_p) : \sigma \in G\}$ forms a $\mathbf{Z}$-basis for $\mathbf{Z}[\zeta_p]$, we deduce that the coefficients n_σ satisfy $n_\sigma \equiv 0 \pmod{q}$ for all $\sigma \in G$. It follows that $\theta \equiv 0 \pmod{q}$ in the ring $\mathbf{Z}[G]$, so that $\psi \equiv 0 \pmod{q}$ is zero in $\mathbf{Z}[G^+]$.

This proves the theorem.

We remark that the binomial coefficient $\binom{2m}{m+1}$ that appears in the preceding proof is equal to m times what is called the mth *Catalan number*. See [45, pp. 219–229].

In terms of the explanation of Runge's method given in chapter 7, the relevant algebraic curve here is given by the equation

$$\prod_{\sigma \in G}(X - \sigma(\zeta_p))^{n_\sigma} = V^q.$$

Since $n_\sigma = n_{\sigma\iota}$ for all $\sigma \in G$, this curve is defined over $\mathbf{Q}(\zeta_p^+)$. From the fact that $m = \frac{1}{q}\sum_\sigma n_\sigma$ is *an integer*, we deduce that the equation looks like

$$X^{mq} + \text{(lower-degree terms)} = V^q.$$

Letting $W = V/X^m$, we see that the divisor at infinity has equation $W^q = 1$. This is the pole divisor of the function X. So, up to Galois conjugacy, the function X has precisely two poles: a pole Q' given by $W = 1$ and a pole Q given by $W = \zeta_q$, where ζ_q denotes a primitive qth root of unity. The pole Q' is defined over $\mathbf{Q}$ and the pole Q is defined over $\mathbf{Q}(\zeta_q)$. The estimates given in the proof of Theorem 12.4 show that the function

$$Z' = V - X^m F_m\left(\frac{1}{X}\right)$$

remains bounded as the point $P = (x, v)$ gets close to the pole Q', i.e., when $|x| \to \infty$. Therefore, Z' has no poles outside Q. The function $Z = DZ'$ with $D = q^{m+\text{ord}q(m!)}$ has the same property and is, in addition, integral over the ring $\mathbf{Z}[X]$. Since the field $\mathbf{Q}(\zeta_p^+)$ is totally real and the pole q is complex, the absolute values of $\phi(Z'(P))$ can be explicitly bounded for all embeddings $\phi : \mathbf{Q}(\zeta_p^+) \hookrightarrow \mathbf{R}$ and, therefore, the integral points on the curve can be determined.

This is what was done in the course of the proof of Theorem 12.4. As in the proof of Cassels' theorem, the main point is to do the estimates uniformly in the exponents p and q.

Exercises

12.1 Let $x \in \mathbf{R}$ and let $k \in \mathbf{Z}_{>0}$. Show the following.

    (a) We have $\left|\binom{x}{k}\right| \le \left|\binom{-|x|}{k}\right|$.

    (b) If $x < 0$, then the binomial coefficient $\binom{x}{k}$ has sign $(-1)^k$.

12.2 Let R be a domain of characteristic 0 with quotient field K and let $I \subset R$ be an ideal. Suppose that the two sequences $\{a_k\}$ and $\{b_k\}$ in R have the property that $a_k \equiv a^k \pmod{I}$ and $b_k \equiv b^k \pmod{I}$ for some $a, b \in R$ and all $k \ge 0$. Show that the product of the two power series $\sum_{k\ge0} \frac{a_k}{k!} T^k$ and $\sum_{k\ge0} \frac{b_k}{k!} T^k$ in $K[[T]]$ is equal to $\sum_{k\ge0} \frac{c_k}{k!} T^k$ with $c_k \in R$ satisfying $c_k \equiv (a+b)^k \pmod{I}$.

12.3 Let p be an odd prime number. Show that the ring of integers of $\mathbf{Q}(\zeta_p^+)$ is equal to $\mathbf{Z}[\zeta_p^+]$. (Hint. see Exercise: 7.2.)

13
Semisimple Group Rings

In this chapter we consider group rings of a finite abelian group with various coefficient rings. The material presented here is well known and can be found in several algebra books.

Let Γ be a finite abelian group and k be a field of characteristic not dividing $\#\Gamma$. Let $\bar{k}$ denote a separable closure of k. Every character $\chi : \Gamma \longrightarrow \bar{k}^*$ takes its values in the subfield of $\bar{k}$ that is generated by the roots of unity of order $\#\Gamma$. The Galois group $\mathrm{Gal}(\bar{k}/k)$ acts on the characters $\chi : \Gamma \longrightarrow k_\Gamma^*$. We let $\chi^\sigma(g) = \sigma(\chi(g))$ for $g \in \Gamma$ and $\sigma \in \mathrm{Gal}(\bar{k}/k)$. For any character $\chi : G \longrightarrow k_\Gamma^*$, we let k_χ denote the subfield of k_Γ generated by the image of χ. We have $k_\chi = k(\zeta_d)$, where ζ_d is a primitive dth root of unity and d is the order of χ. Each field k_χ is a $k[\Gamma]$-algebra via $g \cdot x = \chi(g)x$ for $g \in \Gamma$ and $x \in k_\chi$. In this way, for every character χ, any k_χ-vector space is in a natural way a $k[\Gamma]$-module.

The group ring $k[\Gamma]$ is semisimple. More precisely, there is an isomorphism of $k[\Gamma]$-algebras

$$k[\Gamma] \overset{\cong}{\longrightarrow} \prod_\chi k_\chi.$$

It is the k-linear extension of the homomorphism that maps $g \in \Gamma$ to the vector $(\chi(g))_\chi$. Here χ runs over the $\mathrm{Gal}(\bar{k}/k)$-conjugacy classes of characters $\chi : \Gamma \longrightarrow k_\Gamma^*$.

For any $k[\Gamma]$-module M and character χ, we let $M_\chi = M \otimes_{k[\Gamma]} k_\chi$ denote its χ-part. Taking χ-parts is an exact functor from the category of $k[G]$-modules to the category of k_χ-vector spaces. We have a natural isomorphism of $k[\Gamma]$-modules

$$M \cong \prod_\chi M_\chi.$$

Here χ runs over the $\mathrm{Gal}(\bar{k}/k)$-conjugacy classes of characters $\chi : \Gamma \longrightarrow k_\Gamma^*$.

The following proposition is also valid for non-abelian groups Γ.

Proposition 13.1 *Let Γ be a finite abelian group and let k be a field of characteristic not dividing $\#\Gamma$. Then every exact sequence of $k[\Gamma]$-modules splits.*

R. Schoof, *Catalan's Conjecture*, DOI: 10.1007/978-1-84800-185-5_13,
© Springer-Verlag London Limited 2008

Proof Any short sequence of $k[\Gamma]$-modules

$$0 \longrightarrow A \xrightarrow{f} B \xrightarrow{h} C \longrightarrow 0$$

splits as a sequence of k-vector spaces. We can therefore find a k-linear map $h' : C \longrightarrow B$ with the property that $h \cdot h'$ is the identity map on C. Now we take the "average" over the Γ-conjugates of h'. In other words, consider $h'' : C \longrightarrow B$ given by

$$h''(c) = \frac{1}{\#\Gamma} \sum_{\sigma \in \Gamma} \sigma^{-1}(h'(\sigma(c))).$$

Then h'' is $k[\Gamma]$-linear and $h \cdot h''$ is the identity map on C. The proves the proposition.

Let $k = \mathbf{F}_q$, where q, is a prime number not dividing $\#\Gamma$. For each character $\chi : G \longrightarrow k_\Gamma^*$, let O_χ denote the ring extension of the ring of q-adic integers $\mathbf{Z}_q$ that is generated by the dth roots of unity. Here d is the order of χ. The rings O_χ are discrete valuation rings. In particular, they are principal ideal domains. Since q does not divide $\#\Gamma$, the maximal ideal of O_χ is generated by q. The residue field of O_χ is equal to k_χ. Each ring O_χ is a $\mathbf{Z}_q[\Gamma]$-algebra via $g \cdot x = \chi(g)x$ for $g \in G$ and $x \in O_\chi$. In this way, for every character χ, any O_χ-module is in a natural way a $\mathbf{Z}_q[\Gamma]$-module.

By Exercise 13.2, the map $\operatorname{Hom}(\Gamma, O_\chi^*) \xrightarrow{\cong} \operatorname{Hom}(\Gamma, k_\chi^*)$ induced by the natural homomorphism $O_\chi^* \twoheadrightarrow k_\chi^*$, is a bijection. The Galois group $\operatorname{Gal}(\overline{\mathbf{Q}}_q/\mathbf{Q}_q)$ acts on the characters χ: We have $\chi^\sigma(g) = \sigma(\chi(g))$ for $g \in \Gamma$ and $\sigma \in \operatorname{Gal}(\overline{\mathbf{Q}}_q/\mathbf{Q}_q)$. This action is unramified and is compatible with the action of $\operatorname{Gal}(\overline{\mathbf{F}}_q/\mathbf{F}_q)$ on $\operatorname{Hom}(\Gamma, k_\chi^*)$ that was defined above.

The $\mathbf{Z}_q$-linear extension of the homomorphism that maps $g \in \Gamma$ to the vector $(\chi(g))_\chi$ is an isomorphism of $\mathbf{Z}_q[\Gamma]$-algebras

$$\mathbf{Z}_q[\Gamma] \xrightarrow{\cong} \prod_\chi O_\chi.$$

Here the product runs over the $\operatorname{Gal}(\overline{\mathbf{Q}}_q/\mathbf{Q}_q)$-conjugacy classes of characters χ of Γ. Indeed, since this map is a bijection modulo q, it is surjective by Nakayama's lemma. It is injective because both sides are free modules over $\mathbf{Z}_q$ of the same rank. For any $\mathbf{Z}_q[\Gamma]$-module M, we let $M_\chi = M \otimes_{\mathbf{Z}_q[\Gamma]} O_\chi$ denote its χ-*part*. We have a natural isomorphism of $\mathbf{Z}_q[\Gamma]$-modules

$$M \cong \prod_\chi M_\chi.$$

Here the product runs over the $\operatorname{Gal}(\overline{\mathbf{Q}}_q/\mathbf{Q}_q)$-conjugacy classes of characters χ of Γ.

Proposition 13.2 *Let Γ be a finite abelian group and q be a prime not dividing $\#\Gamma$. Let M be a finite $\mathbf{Z}[\Gamma]$-module. Then the $\mathbf{F}_q[\Gamma]$-modules $M[q]$ and M/qM are isomorphic.*

Proof We may replace M by $M \otimes_{\mathbf{Z}} \mathbf{Z}_q$ and hence assume that it is a $\mathbf{Z}_q[\Gamma]$-module. We consider one χ-part M_χ at the time. Each M_χ is a finite module over the principal ideal domain O_χ. It follows then from the structure theorem of modules over principal ideal domains that $M_\chi[q]$ and M_χ/qM_χ are k_χ-vector spaces of the same dimension. Since we have $\mathbf{F}_q[\Gamma]$-isomorphisms $M[q] \cong \prod_\chi M_\chi[q]$ and $M/qM \cong \prod_\chi M_\chi/qM_\chi$, we see that $M[q]$ and M/qM are isomorphic $\mathbf{F}_q[\Gamma]$-modules, as required.

Lemma 13.3 *Let Γ be a finite abelian group and let q be a prime not dividing $\#\Gamma$. Let M,N be two torsion-free, finitely generated $\mathbf{Z}_q[\Gamma]$-modules. Then M and N are isomorphic if and only if $M \otimes_{\mathbf{Z}_q} \mathbf{Q}_q$ and $N \otimes_{\mathbf{Z}_q} \mathbf{Q}_q$ are isomorphic $\mathbf{Q}_q[\Gamma]$-modules.*

Proof Tensoring the isomorphism $\mathbf{Z}_q[\Gamma] \cong \prod_\chi O_\chi$ with $\mathbf{Q}$, we obtain an isomorphism of $\mathbf{Q}_q[\Gamma]$-algebras:

$$\mathbf{Q}_q[\Gamma] \xrightarrow{\cong} \prod_\chi K_\chi.$$

Here the product runs over the $\mathrm{Gal}(\overline{\mathbf{Q}}_q/\mathbf{Q}_q)$-conjugacy classes of characters χ of Γ and $K_\chi = O_\chi \otimes_{\mathbf{Z}} \mathbf{Q}$ is the quotient field of O_χ. Suppose $M \otimes_{\mathbf{Z}_q} \mathbf{Q}_q$ and $N \otimes_{\mathbf{Z}_q} \mathbf{Q}_q$ are isomorphic as $\mathbf{Q}_q[\Gamma]$-modules. For each character χ, taking χ-parts gives us two free finite-rank O_χ-modules M_χ and N_χ with the property that $M_\chi \otimes_{O_\chi} K_\chi$ and $N_\chi \otimes_{O_\chi} K_\chi$ are isomorphic K_χ-vector spaces. This simply means that they have the same dimension d. Since M and N and hence M_χ and N_χ are torsion-free, it follows that the O_χ-rank of both M_χ and N_χ is also d. This implies that M_χ and N_χ are isomorphic O_χ-modules. Since we have the $\mathbf{Z}_q[\Gamma]$-isomorphisms $M \cong \prod_\chi M_\chi$ and $N \cong \prod_\chi N_\chi$, we see that M and N are isomorphic $\mathbf{Z}_q[\Gamma]$-modules, as required.

Lemma 13.4 *Let Γ be a finite abelian group and let M and N be two finite-dimensional $\mathbf{Q}[\Gamma]$-modules. Then M and N are isomorphic if and only if $M \otimes_{\mathbf{Q}} \mathbf{R}$ and $N \otimes_{\mathbf{Q}} \mathbf{R}$ are isomorphic $\mathbf{R}[\Gamma]$-modules.*

Proof Applying the theory of the beginning of this section to the field $k = \mathbf{Q}$, we obtain a natural isomorphism of $\mathbf{Q}[\Gamma]$-algebras $\mathbf{Q}[\Gamma] \cong \prod_\chi F_\chi$, where F_χ denotes the number field generated by the image of χ.

Tensoring with $\mathbf{R}$, we obtain an isomorphism of $\mathbf{R}[\Gamma]$-algebras:

$$\mathbf{R}[\Gamma] \cong \prod_\chi F_\chi \otimes_{\mathbf{Q}} \mathbf{R}.$$

Here the products run over the $\mathrm{Gal}(\overline{\mathbf{Q}}/\mathbf{Q})$-conjugacy classes of characters χ of Γ. For every character χ, the $\mathbf{R}$-algebra $F_\chi \otimes_{\mathbf{Q}} \mathbf{R}$ is a product of copies of $\mathbf{R}$ and $\mathbf{C}$. If $M \otimes_{\mathbf{Q}} \mathbf{R}$ and $N \otimes_{\mathbf{Q}} \mathbf{R}$ are isomorphic as $\mathbf{R}[\Gamma]$-modules, then for every χ, the

$F_\chi \otimes_\mathbf{Q} \mathbf{R}$-modules $(M \otimes_\mathbf{Q} \mathbf{R}) \otimes_\mathbf{R} (F_\chi \otimes_\mathbf{Q} \mathbf{R})$ and $(N \otimes_\mathbf{Q} \mathbf{R}) \otimes_\mathbf{R} (F_\chi \otimes_\mathbf{Q} \mathbf{R})$ are isomorphic.

For every character $\chi : \Gamma \longrightarrow F_\chi^*$, let M_χ denote the F_χ-module $M \otimes_\mathbf{Q} F_\chi$. For every $\mathbf{Q}[\Gamma]$-module P, we have natural isomorphisms of $F_\chi \otimes_\mathbf{Q} \mathbf{R}$-modules

$$P_\chi \otimes_\mathbf{Q} \mathbf{R} = (P \otimes_\mathbf{Q} \mathbf{R}) \otimes_\mathbf{R} (F_\chi \otimes_\mathbf{Q} \mathbf{R}).$$

We conclude that the $\mathbf{R}[\Gamma]$-modules $M_\chi \otimes_\mathbf{Q} \mathbf{R}$ and $N_\chi \otimes_\mathbf{Q} \mathbf{R}$ are isomorphic. It follows that these two modules have the same $\mathbf{R}$-dimension. It follows that the $\mathbf{Q}$-dimensions and hence the F_χ-dimensions of N_χ and M_χ are equal. Therefore, M_χ and N_χ are isomorphic F_χ-vector spaces. This implies that M and N are isomorphic $\mathbf{Q}[\Gamma]$-modules, as required.

Corollary 13.5 *Let Γ be a finite abelian group and let q be a prime not dividing $\#\Gamma$. Suppose that M and N are $\mathbf{Z}$-free $\mathbf{Z}[G]$-modules and that $M \otimes_\mathbf{Z} \mathbf{R}$ and $N \otimes_\mathbf{Z} \mathbf{R}$ are isomorphic $\mathbf{R}[\Gamma]$-modules. Then M/qM and N/qN are isomorphic $F_q[\Gamma]$-modules.*

Proof Suppose that we have an isomorphism $M \otimes_\mathbf{Z} \mathbf{R} \cong N \otimes_\mathbf{Z} \mathbf{R}$ of $\mathbf{R}[\Gamma]$-modules. Lemma 13.4 implies then that there is an isomorphism $M \otimes_\mathbf{Z} \mathbf{Q} \cong N \otimes_\mathbf{Z} \mathbf{Q}$ of $\mathbf{Q}[\Gamma]$-modules. Tensoring with $\mathbf{Q}_q$ over $\mathbf{Q}$, we obtain an isomorphism $M \otimes_\mathbf{Z} \mathbf{Q}_q \cong N \otimes_\mathbf{Z} \mathbf{Q}_q$ of $\mathbf{Q}_q[\Gamma]$-modules. Since $M \otimes_\mathbf{Z} \mathbf{Z}_q$ and $N \otimes_\mathbf{Z} \mathbf{Z}_q$ are torsion-free, Lemma 13.3 gives us an isomorphism $M \otimes_\mathbf{Z} \mathbf{Z}_q \cong N \otimes_\mathbf{Z} \mathbf{Z}_q$ of $\mathbf{Z}_q[\Gamma]$-modules. The result now follows by reducing this isomorphism modulo q.

We conclude this section with an application to the p-unit group of the cyclotomic field $\mathbf{Q}(\zeta_p)$. First we prove a lemma.

Lemma 13.6 *Let p be an odd prime and let E_p be the group of p-units in $\mathbf{Q}(\zeta_p)$. Let $E_p^+ = E_p \cap \mathbf{Q}(\zeta_p^+)$ and $G^+ = \mathrm{Gal}(\mathbf{Q}(\zeta_p^+)/\mathbf{Q})$. Then there is an an exact sequence of G^+-modules*

$$0 \longrightarrow E_p^+/\{\pm 1\} \longrightarrow E_p/\mu_{2p} \overset{h}{\longrightarrow} \mu_{2p}/\mu_p \longrightarrow 0,$$

Where h is given by $h(u) = u^{\iota-1} = \iota(u)/u$.

Proof For every $u \in E_p$ and for every embedding $\phi : \mathbf{Q}(\zeta_p) \hookrightarrow \mathbf{C}$, the absolute values of the images of $u^{\iota-1}$ under ϕ are equal to 1. Therefore, Exercise 7.1 implies that $h(u)$ is a root of unity. The group of roots of unity of $\mathbf{Q}(\zeta_p)$ is μ_{2p}, and for $u \in \mu_{2p}$, the number $u^{\iota-1} = u^{-2}$ is contained in μ_p. Therefore, the map h is a well-defined group homomorphism. It follows that the G-module E_p/μ_{2p} is ι-invariant and hence has the structure of a G^+-module. The surjectivity of h follows from the fact that $h(\zeta_p - 1) = (\zeta_p - 1)^{\iota-1} = -\zeta_p^{-1}$. If we have $h(u) = u^{\iota-1} \in \mu_p$, then $(u\zeta)^{\iota-1} = 1$ for some $\zeta \in \mu_{2p}$. In other words, $u = \zeta u'$ for some $u' \in E_p^+$. Finally, we have $\mu_{2p} \cap E_p^+ = \{\pm 1\}$. These facts show that the sequence is exact. This proves the lemma.

Proposition 13.7 *Let p be an odd prime and $G^+ = \mathrm{Gal}(Q(\zeta_p^+)/Q)$. Suppose that q is a prime not dividing $p(p-1)$. Then the p-unit group E_p of $Q(\zeta_p)$ has the property that E_p/E_p^q is free of rank 1 over $F_q[G^+]$.*

Proof Let $E_p^+ = E_p \cap Q(\zeta_p)$. Since p is totally ramified, there lies only one prime over p and it follows from Dirichlet's unit theorem that the logarithmic map

$$\lambda : E_p^+ \hookrightarrow \prod_\phi R$$

given by $\varepsilon \mapsto (\log|\phi(\varepsilon)|)_\phi$ maps $E_p^+/\{\pm 1\}$ isomorphically to a lattice of rank $\#G^+$ inside $\prod_\phi R$. Here the products run over the set of embeddings $\phi : Q(\zeta_p^+) \hookrightarrow R$.

The logarithmic map is easily seen to be $Z[G^+]$-linear. It follows that the $R[G^+]$-module $(E_p^+/\{\pm 1\}) \otimes R$ is isomorphic to $\prod_\phi R$.

By Exercise 13.3, the $R[G^+]$-module

$$Q(\zeta_p^+) \otimes_Q R \cong \prod_\phi R$$

is free of rank 1. Therefore, the G^+-modules $E_p^+/\{\pm 1\}$ and $Z[G^+]$ become isomorphic when we tensor with them R. By Lemma 13.6, the G^+-modules $E_p^+/\{\pm 1\}$ and E_p/μ_{2p} also become isomorphic when we tensor with R. Since all modules are Z-free and we have $\mu_{2p} \subset E_p^q$, the result follows from Corollary 13.5.

Exercises

13.1 Let Γ be a finite group and let k be a field whose characteristic does not divide $\#\Gamma$. For any character of Γ, let k_χ denote the extension of k that is generated by the values of χ. Show that

$$\sum_\chi [k_\chi : k] = \#\Gamma.$$

Here the sum runs over the $\mathrm{Gal}(\overline{k}/k)$-conjugacy classes of the characters of Γ.

13.2 Let Γ be a finite group and q be a prime not dividing the order of Γ. For every character χ of Γ, let O_χ denote the Z_q-algebra that is generated by the image of Γ. Let k_χ denote the residue field of the local ring O_χ. Show that the reduction homomorphism

$$\mathrm{Hom}(\Gamma, O_\chi^*) \longrightarrow \mathrm{Hom}(\Gamma, k_\chi^*)$$

is an isomorphism of groups.

13.3 Let K be a finite Galois extension of $\mathbf{Q}$ with $H = \text{Gal}(K/\mathbf{Q})$. Suppose that K is a totally real number field. We let the group H act on

$$\prod_\phi \mathbf{R}$$

via $\sigma(x_\phi)_\phi = (x_{\phi \cdot \sigma^{-1}})_\phi$. Here the product runs over the various embeddings $\phi : K \hookrightarrow \mathbf{R}$. In this way, $\prod_\phi \mathbf{R}$ acquires the structure of an $\mathbf{R}[H]$-module.

(a) Show that $\prod_\phi \mathbf{R}$ is a free $\mathbf{R}[H]$-module of rank 1.
(b) Show that the natural map $K \longrightarrow \prod_\phi \mathbf{R}$ given by $x \mapsto (\phi(x))_\phi$ induces an isomorphism $K \otimes \mathbf{R} \longrightarrow \prod_\phi \mathbf{R}$ of $\mathbf{R}[H]$-modules.

13.4 Let p be a prime and let Γ be a group of order p.

(a) Give an example of a *nonsplit* exact sequence of $\mathbf{F}_p[\Gamma]$-modules.
(b) Give an example of a finite $\mathbf{Z}_p[\Gamma]$-module M with the property that the $\mathbf{F}_p[\Gamma]$-modules $M[p]$ and M/pM are *not* isomorphic.

13.5 Show, by means of a computer calculation, that for $p = 1129$ and $q = 3$, the group E_p/E_p^q is *not* a free $\mathbf{F}_q[G^+]$-module. See [42].

14
The Plus Argument II

In this chapter, we prove Theorem II of chapter 1. The key ingredients are the results of chapter 12 and Francisco Thaine's famous theorem [48], which is proved in chapter 16. Let p, q be odd primes and let $x, y \in \mathbf{Z}$ be a nonzero solution to Catalan's equation $x^p - y^q = 1$. By Lemma 6.1, we have $p \neq q$. We use the notation introduced in Chapter 7. In particular, we let $G = \mathrm{Gal}(\mathbf{Q}(\zeta_p)/\mathbf{Q})$, we write ι for the complex conjugation σ_{-1}, and we let $G^+ = G/\langle\iota\rangle$.

In this chapter, we write E for the subgroup E_p of p-units of $\mathbf{Q}(\zeta_p)^*$. Similarly, we write C for the the group C_p of *cyclotomic* p-units. The group C_p is the multiplicative $\mathbf{Z}[G]$-module generated by $1 - \zeta_p$. See Exercise 14.4.

The obstruction group

$$H = \{\alpha \in \mathbf{Q}(\zeta_p)^* : \mathrm{ord}_{\mathfrak{r}}(\alpha) \equiv 0 \ (\mathrm{mod}\ q) \text{ for all primes } \mathfrak{r} \neq \mathfrak{p}\}/\mathbf{Q}(\zeta_p)^{*q}.$$

was introduced in Chapter 7. It is an $\mathbf{F}_q[G]$-module. We consider the Selmer group S of Chapter 10. It is the $\mathbf{F}_q[G]$-submodule of H defined by

$$S = \{\alpha \in H : \alpha \text{ is a } q\text{-adic } q\text{th power}\}.$$

Here we call $\alpha \in \mathbf{Q}(\zeta_p)^*$ a "q-adic q-th power", if it is a qth power in the completion $F_{\mathfrak{q}}$ of $\mathbf{Q}(\zeta_p)$ at each of the primes $\mathfrak{q}$ lying over q.

By Corollary 10.3, the number $x - \zeta_p$ is contained in S. Therefore, $(x - \zeta_p)^{1+\iota}$ is contained in the plus part S^+ of S. In this chapter, we show that S^+ is annihilated by some nonzero element of $\mathbf{F}_q[G^+]$. This then contradicts the fact, proved in chapter 12, that $(x - \zeta_p)^{1+\iota}$ generates a *free* $\mathbf{F}_q[G^+]$-module inside H^+. We prove our annihilation result under the assumption that $p \not\equiv 1 \ (\mathrm{mod}\ q)$. In this case, q does not divide $\#G^+$ and the group ring $\mathbf{F}_q[G^+]$ is *semisimple*. Therefore, the theory of chapter 13 applies and the results there are used in several places in the proof.

Theorem 14.1 *Suppose that $p > q$ are odd primes and that $p \not\equiv 1 \ (\mathrm{mod}\ q)$. Then the $\mathbf{F}_q[G^+]$-module S^+ has a nonzero annihilator.*

Proof By $E^{(q)}$ we denote the subgroup $\{\varepsilon \in E : \varepsilon \text{ is a } q\text{-adic } q\text{th power}\}$ of the group E of p-units. It contains the subgroup E^q of qth powers. The exact sequence of Proposition 7.4 gives rise to the exact sequence

R. Schoof, *Catalan's Conjecture*, DOI: 10.1007/978-1-84800-185-5_14,
© Springer-Verlag London Limited 2008

$$0 \longrightarrow \quad E^{(q)}/E^q \quad \longrightarrow \quad S^+ \quad \longrightarrow \quad Cl_p^+[q].$$

By $C^{(q)}$ we denote the subgroup of the group C of cyclotomic p-units that consists of the cyclotomic p-units that are q-adic qth powers. It is a $\mathbf{Z}[G]$-module. By Lemma 7.1, the group E/E^q is an $\mathbf{F}_q[G^+]$-module. It admits a filtration

$$0 \underbrace{\subset}_{E_1} C^{(q)}E^q/E^q \underbrace{\subset}_{E_2} CE^q/E^q \underbrace{\subset}_{E_3} E/E^q$$

with successive subquotients $E_1 = C^{(q)}E^q/E^q$, $E_2 = CE^q/C^{(q)}E^q$, and $E_3 = E/CE^q$, respectively. The natural map $E_1 \hookrightarrow E^{(q)}/E^q$ is injective. Its cokernel is $E^{(q)}/C^{(q)}E^q$. Since $E^{(q)} \cap CE^q = C^{(q)}E^q$, the natural map $E^{(q)}/C^{(q)}E^q \longrightarrow E/CE^q$ is injective. Since $E/CE^q = E_3$, we obtain the following diagram with exact rows and columns:

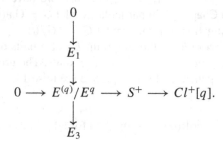

Since q does not divide $\#G^+$, Proposition 13.1 applies and every exact sequence of $\mathbf{F}_q[G^+]$-modules splits. Therefore, we have an injection of $\mathbf{F}_q[G^+]$-modules:

$$S^+ \quad \hookrightarrow E_1 \times E_3 \times Cl^+[q].$$

Theorem 16.3 below is a special case of Thaine's theorem [48]. It says that the $\mathbf{F}_q[G^+]$-annihilator of $E_3 = E/CE^q$ *also* annihilates Cl^+/Cl^{+q}. Since q does not divide $\#G^+$, Proposition 13.2 implies that the $\mathbf{F}_q[G^+]$-modules Cl^+/Cl^{+q} and $Cl^+[q]$ are isomorphic. Therefore, the $\mathbf{F}_q[G^+]$-annihilator of $E_3 = E/CE^q$ annihilates $Cl^+[q]$. It follows that the $\mathbf{F}_q[G^+]$-annihilator of $E_3 \times E_1$ annihilates $E_1 \times E_3 \times Cl^+[q]$ and hence S^+. Therefore, we are done if we show that the $\mathbf{F}_q[G^+]$-annihilator of $E_3 \times E_1$ is *not* zero.

Suppose it is. By Proposition 13.1, all exact sequences of $\mathbf{F}_q[G^+]$-modules split. Therefore, $E_1 \times E_2 \times E_3$ is isomorphic to E/E^q. By Corollary 13.7, the $\mathbf{F}_q[G^+]$-module E/E^q is free of rank 1. The fact that the $\mathbf{F}_q[G^+]$-annihilator of $E_3 \times E_1$ is zero implies by Exercise 14.1 that the annihilator of E_2 is the unit ideal. This implies $E_2 = 0$. Since $E_2 = CE^q/C^{(q)}E^q$, this means that every cyclotomic p-unit is a q-adic qth power. Since $p > q$, this is impossible by Proposition 14.2 below.

This proves the theorem.

Let p, q be odd primes. When $p \neq 3$, it is highly unlikely that all cyclotomic p-units in $\mathbf{Q}(\zeta_p)$ are q-adic q-th powers. At present, however, this can only by shown when $p \geq q$.

Proposition 14.2 *Let p, q be odd primes. If every cyclotomic p-unit contained in $\mathbf{Q}(\zeta_p)$ is a q-adic qth power, then one has $p < q$.*

Proof Let ζ be a primitive pth root of unity in $\mathbf{Q}(\zeta_p)$. If every cyclotomic p-unit is a q-adic qth power, then every cyclotomic p-unit is congruent to a qth power modulo q^2. We have in particular

$$\zeta^q - 1 \equiv u^q \ (\mathrm{mod} \ q^2) \quad \text{for some} \ u \in \mathbf{Q}(\zeta_p)^*.$$

From the fact that $(\zeta - 1)^q \equiv \zeta^q - 1 \ (\mathrm{mod} \ q)$, we deduce that

$$(\zeta - 1)^q \equiv u^q \ (\mathrm{mod} \ q).$$

Both sides of the equations are qth powers. By Exercise 10.2, the ring $\mathbf{Z}[\zeta_p]/(q)$ contains no non-zero nilpotent elements, so that Exercise 10.1 implies that the congruence also holds modulo q^2:

$$(\zeta - 1)^q \equiv \zeta^q - 1 \ (\mathrm{mod} \ q^2).$$

This means that the pth root of unity ζ is a zero of the "Witt polynomial":

$$W(T) = \frac{(T-1)^q - T^q + 1}{qT} \in \mathbf{F}_q[T].$$

This implies that in the ring $\mathbf{F}_q[T]$, the polynomial $W(T)$ is divisible by the pth cyclotomic polynomial. Inspection of the degrees of these polynomials shows that necessarily $p - 1 \leq q - 2$, as required.

This proves the proposition.

Now we prove Theorem II of Chapter 1.

Theorem II *Let p, q be odd primes and suppose that x, y are nonzero integers satisfying Catalan's equation $x^p - y^q = 1$. Then we have either $q \equiv 1 \ (\mathrm{mod} \ p)$ or $p \equiv 1 \ (\mathrm{mod} \ q)$.*

Proof By Lemma 6.1, we have $p \neq q$. By symmetry, we may assume that $p > q$. Then, clearly, $q \not\equiv 1 \ (\mathrm{mod} \ p)$ holds. Suppose that also $p \not\equiv 1 \ (\mathrm{mod} \ q)$.

Theorem IV implies that we may assume $p, q \geq 7$ and that therefore we may apply Theorem 12.4. It follows that the $\mathbf{F}_q[G^+]$-submodule of H^+ generated by $(x - \zeta_p)^{1+\iota}$ is free. On the other hand, by Corollary 10.3, the element $(x - \zeta_p)^{1+\iota}$ is contained in the group S^+ and Theorem 14.1 says that S^+ admits a nontrivial $\mathbf{F}_q[G^+]$-annihilator. This contradiction implies Theorem II.

The proof of Theorem II uses Theorem IV only when one of p, q is smaller than 7. Therefore, one needs only to check the conditions of Theorem IV for the primes 3 and 5 only. In other words, it suffices to know that the minus class numbers of the cyclotomic fields $\mathbf{Q}(\zeta_3)$ and $\mathbf{Q}(\zeta_5)$ are both equal to 1.

Exercises

14.1 Let R be a ring and let A, B be two R-modules with the property that $A \times B$ is a free R-module of rank 1. Show that the R-annihilators of A and B are coprime ideals. In other words, show that $\mathrm{Ann}_R(A) + \mathrm{Ann}_R(B) = R$.

14.2 Let q be a prime. Show that the polynomial $W(T)$ of Proposition 14.2 is congruent to

$$\sum_{j=0}^{q-2} \frac{T^j}{j+1} \ (\mathrm{mod} \ q).$$

14.3 Show that $T^2 - T + 1$ divides the polynomial $W(T) = ((T-1)^q - T^q + 1)/qT$ for every prime $q \geq 5$.

14.4 Show that the group of cyclotomic p-units $C \subset \mathbf{Q}(\zeta_p)^*$ contains μ_{2p}. Show that C/μ_{2p} is a free $\mathbf{Z}[G^+]$-module of rank 1 generated by $1 - \zeta_p$. Here G^+ denotes the Galois group of $\mathbf{Q}(\zeta_p^+)$ over $\mathbf{Q}$.

15
The Density Theorem

In this chapter, we prove N.G. Chebotarëv's density theorem (Fig. 15.1). This important result regards densities of certain sets of prime ideals of a given number field. Dirichlet's theorem on primes in arithmetic progressions is a special case. Our proof follows Chebotarëv's original strategy. This method does not involve any class field theory. The results of this section are classical and can be found in several textbooks, e.g. [24]. See also [28].

Let F be a number field and let K be a finite Galois extension of F. Let $G = \text{Gal}(K/F)$. For any prime ideal $\mathfrak{q}$ of K, we define its *decomposition group* by $G_{\mathfrak{q}} = \{\sigma \in G : \sigma(\mathfrak{q}) = \mathfrak{q}\}$. Let $\mathfrak{p} = \mathfrak{p} \cap O_F$ denote the prime ideal of F over which $\mathfrak{q}$ lies. The natural homomorphism from $G_{\mathfrak{p}}$ to the Galois group of $O_K/\mathfrak{q}$ over $O_F/\mathfrak{p}$ is a well-defined surjective homomorphism. If $\mathfrak{p}$ is *unramified* in $K \subset L$, this homomorphism is also injective. The Galois group of $O_K/\mathfrak{q}$ over $O_F/\mathfrak{p}$ is cyclic. It is generated by the *Frobenius automorphism* of the finite field $O_K/\mathfrak{q}$. This is the automorphism given by

$$x \mapsto x^{N(\mathfrak{p})} \quad \text{for } x \in O_K/\mathfrak{q}.$$

Here $N(\mathfrak{p}) = \#(O_F/\mathfrak{p})$. The fixed field of the Frobenius automorphism is precisely the subfield $O_F/\mathfrak{p}$.

Definition Let $\mathfrak{p}$ be a prime ideal of F that is unramified in the extension $F \subset K$. Let $\mathfrak{q}$ be a prime of K that lies over $\mathfrak{p}$. The *Frobenius element* $\varphi_{\mathfrak{q}}$ of $\mathfrak{q}$ is the unique automorphism in $G_{\mathfrak{q}}$ that induces the Frobenius automorphism of the finite residue field $O_K/\mathfrak{q}$.

Since the primes $\mathfrak{q}$ lying over $\mathfrak{p}$ are Galois conjugates of one another, the corresponding Frobenius elements $\varphi_{\mathfrak{q}}$ are conjugate in the Galois group G. In fact, the Frobenius elements constitute a full conjugacy class of the group G. This conjugacy class depends only on the prime ideal $\mathfrak{p}$, and we call it the *Frobenius conjugacy class associated to* $\mathfrak{p}$. The Frobenius conjugacy class of $\mathfrak{p}$ is the identity if and only if $\mathfrak{p}$ is totally split in K.

R. Schoof, *Catalan's Conjecture*, DOI: 10.1007/978-1-84800-185-5_15,
© Springer-Verlag London Limited 2008

Fig. 15.1 N. G. Chebotarëv (1894–1947). (Reproduced by with the kind permission of the Museum of the History of Kazan University, a division of Kazan State University.)

In the special case where G is *abelian*, the conjugacy class consists of one single element. It is the Frobenius element of any of the primes q that lie over p. In this case, we may also call φ_q the *Frobenius element of* p.

We look in more detail at two special situations.

Example 15.1 (*Cyclotomic extensions*) The homomorphism

$$(\mathbf{Z}/m\mathbf{Z})^* \longrightarrow \mathrm{Gal}(\mathbf{Q}(\zeta_m)/\mathbf{Q})$$

given by $x \mapsto \sigma_x$ is an isomorphism. Here σ_x is the automorphism given by the formula $\sigma_x(\zeta_m) = \zeta_m^x$. The primes p that do not divide m are unramified in the extension $\mathbf{Q} \subset \mathbf{Q}(\zeta_m)$. The Frobenius element φ_q of any prime q lying over an unramified prime p is the element σ_p of the group $\mathrm{Gal}(\mathbf{Q}(\zeta_m)/\mathbf{Q})$. It corresponds to the congruence class of p in $(\mathbf{Z}/m\mathbf{Z})^*$. Since $\mathrm{Gal}(\mathbf{Q}(\zeta_m)/\mathbf{Q})$ is abelian, the Frobenius element depends only on p and not on the prime q lying over p.

More generally, let F be a number field and m be an integer. Then all prime ideals p that do not divide m are unramified in the extension $F \subset F(\zeta_m)$. The restriction map is an injective homomorphism from the Galois group $\mathrm{Gal}(F(\zeta_m)/F)$ into $\mathrm{Gal}(\mathbf{Q}(\zeta_m)/\mathbf{Q})$. It is an isomorphism if and only if $F \cap \mathbf{Q}(\zeta_m) = \mathbf{Q}$. The Frobenius element of an unramified prime q of $F(\zeta_m)$ lying over the prime p of F is the element $\sigma_{N(\mathfrak{p})}$. In terms of the isomorphism $\mathrm{Gal}(\mathbf{Q}(\zeta_m)/\mathbf{Q}) \cong (\mathbf{Z}/m\mathbf{Z})^*$, the Frobenius element is the congruence class of $N(\mathfrak{p})$ in $(\mathbf{Z}/m\mathbf{Z})^*$. As in the case $F = \mathbf{Q}$, the Frobenius element depends only on the prime p and not on the primes q lying over it.

Example 15.2 (*Kummer extensions*) Let m be a natural number and let F be a number field containing the mth roots of unity. For any finitely generated subgroup A of F^*, we write $F(\sqrt[m]{A})$ for the field extension of F that is generated by the mth roots of the elements of A. The field $F(\sqrt[m]{A})$ is a *finite* extension of F. The natural *Kummer homomorphism*

$$\mathrm{Gal}(F(\sqrt[m]{A})/F) \longrightarrow \mathrm{Hom}(A/A \cap (F^*)^m, \mu_m)$$

given by $\sigma \mapsto f_\sigma$, where $f_\sigma(a) = \sigma(\sqrt[m]{a})/\sqrt[m]{a}$ for $a \in A$, is a well-defined isomorphism of groups. See Exercise 9.9.

The prime ideals $\mathfrak{p}$ of F that do not divide m or any of the numerators and denominators of the elements in A, are unramified in the extension $F \subset F(\sqrt[q]{A})$. By Exercise 15.1, the number $N(\mathfrak{p}) - 1$ is divisible by m. Since $\mathrm{Gal}(F(\sqrt[m]{A})/F)$ is commutative, the Frobenius elements $\varphi_\mathfrak{q}$ of the primes $\mathfrak{q}$ lying over $\mathfrak{p}$ are equal to one another. They are equal to the homomorphism $A/(A \cap F^{*m}) \longrightarrow \mu_m$ that maps $a \in A$ to the unique root of unity $\xi \in \mu_m$ for which

$$\xi \equiv a^{\frac{N(\mathfrak{p})-1}{m}} \pmod{\mathfrak{p}}.$$

Indeed, by definition of the Kummer homomorphism, we have

$$a^{\frac{N(\mathfrak{p})-1}{m}} \equiv \frac{\varphi_\mathfrak{q}(\sqrt[m]{a})}{\sqrt[m]{a}} \pmod{\mathfrak{p}}$$

for all $a \in A$. This shows that

$$\varphi_\mathfrak{q}(\sqrt[m]{a}) = \sqrt[m]{a}\,\xi \equiv \sqrt[m]{a} \cdot a^{\frac{N(\mathfrak{p})-1}{m}} \equiv \sqrt[m]{a}^{N(\mathfrak{p})} \pmod{\mathfrak{q}}$$

for any prime $\mathfrak{q}$ lying over $\mathfrak{p}$, as required.

The zeta function $\zeta_F(s)$ of a number field F admits an Euler product. We have

$$\zeta_F(s) = \prod_\mathfrak{p} \left(1 - \frac{1}{N(\mathfrak{p})^s}\right)^{-1} \qquad \text{for } s \in \mathbf{C} \text{ with } \mathrm{Re}(s) > 1,$$

where the product runs over the prime ideals of F. The zeta function $\zeta_F(s)$ admits a meromorphic extension to $\mathbf{C}$ with a simple pole at $s = 1$. See [24]. Therefore, Exercise 15.3 implies that for $s \in \mathbf{R}_{>1}$, we have

$$-\log(s-1) = -\sum_\mathfrak{p} \log\left(1 - \frac{1}{N(\mathfrak{p})^s}\right)$$

$$= \sum_\mathfrak{p} \frac{1}{N(\mathfrak{p})^s} + g(s),$$

where $g(s)$ is a function that remains bounded as s approaches 1. Writing $h_1(s) \sim h_2(s)$ when two functions $h_1, h_2 : \mathbf{R}_{>1} \longrightarrow \mathbf{C}$ have the property that $h_1(s) - h_2(s)$ remains bounded as $s \to 1$, we have

$$\sum_{\mathfrak{p}} \frac{1}{N(\mathfrak{p})^s} \sim \log\left(\frac{1}{s-1}\right).$$

In particular, we have

$$\lim_{s \downarrow 1} \sum_{\mathfrak{p}} \frac{1}{N(\mathfrak{p})^s} = +\infty.$$

Definition   Let F be a number field and S be a set of prime ideals of F. The *lower density* $d_F^-(S)$ and *upper density* $d_F^+(S)$ of S are defined by

$$d_F^-(S) = \liminf_{s \downarrow 1} \frac{\sum_{\mathfrak{p} \in S} \frac{1}{N(\mathfrak{p})^s}}{\log(\frac{1}{s-1})}$$

and

$$d_F^+(S) = \limsup_{s \downarrow 1} \frac{\sum_{\mathfrak{p} \in S} \frac{1}{N(\mathfrak{p})^s}}{\log(\frac{1}{s-1})},$$

respectively. We have

$$0 \leq d_F^-(S) \leq d_F^+(S) \leq 1.$$

If the lower density $d_F^-(S)$ is equal to the upper density $d_F^+(S)$, we define the *density* $d_F(S)$ by

$$d_F(S) = d_F^-(S) = d_F^+(S).$$

As we explained, the denominator in the definitions of $d_F^+(S)$ and $d_F^-(S)$ is, up to a bounded function, equal to $\sum_{\mathfrak{p}} \frac{1}{N(\mathfrak{p})^s}$ where the sum runs over *all* primes $\mathfrak{p}$ of F. Therefore, the quantity $d_F(S)$ is called a *density*. It measures, in a certain sense, the proportion of primes of F that is in S. Finite sets S have density zero. Exercise 15.6 implies that for any set S, the set that consists of the prime ideals in S *of degree 1* has the same upper and lower density as S.

Although we do not need it here, we mention that the *natural density* $d_F^{\mathrm{nat}}(S)$ of a set of primes S is defined as

$$d_F^{\mathrm{nat}}(S) = \lim_{X \to \infty} \frac{\#\{\mathfrak{p} \in S : \mathfrak{p} \leq X\}}{\#\{\mathfrak{p} : \mathfrak{p} \leq X\}}.$$

The notion of natural density is perhaps closer to our intuitive notion of *density*. The following proposition explains the relation to our notion of density. It is not needed for the proof of Chebotarëv's density theorem.

Proposition 15.3 *Let F be a number field and let S be a set of prime ideals of F. If the natural density $d_F^{nat}(S)$ exists, then so does the density $d_F(S)$. In addition, they are equal.*

Proof For any $X \in \mathbf{R}$, let $\pi(X)$ denote the number of prime ideals of F of norm $< X$ and let $\pi_S(X)$ denote the number of prime ideals in S of norm $< X$. We have

$$d_F^{nat}(S) = \lim_{X \to \infty} \frac{\pi_S(X)}{\pi(X)}.$$

We write $d = d_F^{nat}(S)$ and let $\varepsilon > 0$. Then there exists X_0 so that for all $X > X_0$, we have

$$|\pi_S(x) - d\pi(X)| < \varepsilon\pi(X).$$

Let $s \in \mathbf{R}_{s>1}$. We have $\sum_{\mathfrak{p}} \frac{1}{N(\mathfrak{p})^s} = \sum_{n \geq 1} \pi(n)(\frac{1}{n^s} - \frac{1}{(n+1)^s})$ and, similarly, $\sum_{\mathfrak{p} \in S} \frac{1}{N(\mathfrak{p})^s} = \sum_{n \geq 1} \pi_S(n)(\frac{1}{n^s} - \frac{1}{(n+1)^s}))$. It follows that

$$\left| \sum_{\mathfrak{p} \in S} \frac{1}{N(\mathfrak{p})^s} - d \sum_{\mathfrak{p}} \frac{1}{N(\mathfrak{p})^s} \right| = \left| \sum_{n \geq 1} (\pi_S(n) - d\pi(n)) \left(\frac{1}{n^s} - \frac{1}{(n+1)^s} \right) \right|.$$

We split the sum over n into two parts: a sum over $n \leq X_0$ and one over $n > X_0$. This leads to

$$\left| \sum_{\mathfrak{p} \in S} \frac{1}{N(\mathfrak{p})^s} - d \sum_{\mathfrak{p}} \frac{1}{N(\mathfrak{p})^s} \right| \leq C(X_0) + \varepsilon \sum_{\mathfrak{p}} \frac{1}{N(\mathfrak{p})^s}.$$

Finally, we divide by $\sum_{\mathfrak{p}} \frac{1}{N(\mathfrak{p})^s}$ and let S tend to 1. We find

$$\left| \lim_{s \downarrow 1} \frac{\sum_{\mathfrak{p} \in S} \frac{1}{N(\mathfrak{p})^s}}{\sum_{\mathfrak{p}} \frac{1}{N(\mathfrak{p})^s}} - d \right| \leq \varepsilon.$$

Since this inequality holds for every $\varepsilon > 0$, the result follows.

Our proof of Chebotarëv's density theorem starts off by proving a special case. It is the generalization of Dirichlet's theorem on primes in arithmetic progressions to an arbitrary number field F.

Proposition 15.4 *Let F be a number field and m be a natural number. Let H be the Galois group of $F(\zeta_m)$ over F. Then for every $h \in H$, the density of the set of prime* *ideals $\mathfrak{p}$ of F for which the Frobenius element is h, is equal to $1/\#H$. In particular, for every $h \in H$, the set of such prime ideals is infinite.*

Proof The restriction map identifies the Galois group of $F(\zeta_m)$ over F with a subgroup H of $(\mathbf{Z}/m\mathbf{Z})^* \cong \mathrm{Gal}(\mathbf{Q}(\zeta_m)/\mathbf{Q})$. For every prime ideal $\mathfrak{p}$ of F, its Frobenius element is $N(\mathfrak{p}) \in H \subset (\mathbf{Z}/m\mathbf{Z})^*$. Here the norm is taken from F to $\mathbf{Q}$.

For every character $\chi : H \longrightarrow \mathbf{C}^*$, we define the *Dirichlet L-series* $L(s, \chi)$ by

$$L(s, \chi) = \sum_{I \subset O_F} \frac{\chi(N(I))}{N(I)^s}$$

$$= \prod_{\mathfrak{p}} \left(1 - \frac{\chi(N(\mathfrak{p}))}{N(\mathfrak{p})^s} \right)^{-1}$$

for $s \in \mathbf{C}$ with $\mathrm{Re}(s) > 1$. Here the sum runs over all nonzero O_F-ideals I and the product runs over the primes $\mathfrak{p}$ of F. By convention, we put $\chi(k) = 0$ whenever k is an integer that is not prime to the conductor of χ. See Exercise 15.7. The fact that the L-series $L(s, \chi)$ admits Euler product expansions follows from the fact that every fractional F-ideal is in a unique way a product of prime ideals. We have

$$\zeta_{F(\zeta_m)}(s) = \prod_{\chi} L(s, \chi),$$

where the product runs over the characters χ of H. The L-series corresponding to the trivial character is precisely the zeta function of the field F. By Exercise 15.8, the L-series corresponding to the nontrivial characters converge for $s \in \mathbf{R}_{>1}$. Since both zeta functions $\zeta_F(s)$ and $\zeta_{F(\zeta_m)}(s)$ have simple poles of order 1 at $s = 1$, we have

$$L(1, \chi) \neq 0 \qquad \text{for each nontrivial character } \chi.$$

Taking logarithms, we see that for each nontrivial character χ, the sum

$$\sum_{\mathfrak{p}} \frac{\chi(N(\mathfrak{p}))}{N(\mathfrak{p})^s}$$

remains bounded as $s \to 1$.

Finally, for each $h \in H$, we consider the functions $f_h(s)$ on $\mathbf{R}_{s>1}$ given by

$$f_h(s) = \sum_{\chi} \chi^{-1}(h) \sum_{\mathfrak{p}} \frac{\chi(N(\mathfrak{p}))}{N(\mathfrak{p})^s} \qquad (s \in \mathbf{C}, \mathrm{Re}(s) > 1).$$

Here the sum runs over the characters of H. Up to a function that remains bounded as $s \to 1$, the function $f_h(s)$ is equal to the L-series corresponding to $\chi = 1$. Since the latter function is the zeta function of F, we have

$$f_h(s) \sim \log\left(\frac{1}{s-1}\right).$$

Changing the order of summation, we find

$$f_h(s) \sim \sum_{\mathfrak{p}} \frac{\sum_\chi \chi(N(\mathfrak{p})h^{-1})}{N(\mathfrak{p})^s},$$

which, by the orthogonality relations for characters, becomes

$$\log\left(\frac{1}{s-1}\right) \sim \sum_{\substack{\mathfrak{p} \\ N(\mathfrak{p})\equiv h \ (\mathrm{mod}\ m)}} \frac{\#H}{N(\mathfrak{p})^s}.$$

It follows that the density of the set of prime ideals $\mathfrak{p}$ satisfying $N(\mathfrak{p}) \equiv h$ (mod m) is $1/\#H$, as required.

The key point is to prove Chebotarëv's density theorem for *cyclic* extensions. The next proposition takes care of this case.

Proposition 15.5 *Let F be a number field and $F \subset K$ be a cyclic Galois extension. Put $G = \mathrm{Gal}(K/F)$. Then for every $\sigma \in G$, the set S of prime ideals $\mathfrak{p}$ of F with Frobenius element equal to σ has a density. It is given by*

$$d_F(S) = \frac{1}{\#G}.$$

Proof Let $\sigma \in G$ and q be any prime number that is not ramified in the extension $\mathbf{Q} \subset K$. Let ζ denote a primitive qth root of unity. Since $K \cap \mathbf{Q}(\zeta) = \mathbf{Q}$, the Galois group H of $F(\zeta)$ over F is isomorphic to $(\mathbf{Z}/q\mathbf{Z})^*$, and the restriction map from $\mathrm{Gal}(K(\zeta)/F)$ to $\mathrm{Gal}(K/F) \times \mathrm{Gal}(F(\zeta)/F)$ identifies $\mathrm{Gal}(K(\zeta)/F)$ with $G \times H$.

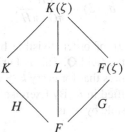

Let S be the set of prime ideals of F whose Frobenius element is equal to σ in G. Then we have

$$S = \bigcup_{\tau \in H} S_\tau,$$

where S_τ denotes the set of prime ideals of F whose Frobenius element is equal to $(\sigma, \tau) \in G \times H$. Since at this point in the proof we do not know that the sets S and S_τ actually have well-defined densities, we work with *lower* densities. The latter always exist. By Exercise 15.4, we have

$$d_F^-(S) \geq \sum_{\tau \in H} d_F^-(S_\tau).$$

Now let $n = \#G$ and suppose that the prime q satisfies $q \equiv 1 \pmod{n}$. Fix $\tau \in H \cong (\mathbf{Z}/q\mathbf{Z})^*$ of order divisible by n. Then the subgroups $\langle(\sigma, \tau)\rangle$ and $G \times \{1\}$ of $G \times H$ have trivial intersection. Therefore, the composite of the corresponding fields of invariants is equal to $K(\zeta)$. Denoting the invariant field of $\langle(\sigma, \tau)\rangle$ by L, this means precisely that $L(\zeta) = K(\zeta)$, so that the extension $L \subset K(\zeta)$ is cyclotomic. By Proposition 15.3, the density of the set of primes T of L with Frobenius element in $\mathrm{Gal}(K(\zeta)/L) \subset G \times H$ equal to (σ, τ) exists; it is given by

$$d_L(T) = \frac{1}{[K(\zeta) : L]}.$$

The subset of T consisting of the primes of degree 1 over L has the same density. Therefore, we have $d_L(T) = d_L^-(T) = [L : F]d_F^-(S_\tau)$. It follows that the density of the set S_τ of primes $\mathfrak{p}$ of F with Frobenius element in $G \times H$ equal to (σ, τ) is given by

$$d_F^-(S_\tau) = \frac{1}{[L : F][K(\zeta) : L]} = \frac{1}{\#G \cdot \#H}.$$

Summing over all automorphisms $\tau \in H$ of order divisible by n, one obtains an estimate for the *lower density* of the set S. It is given by

$$d_F^-(S) \geq \frac{\#H_n}{\#G \cdot \#H},$$

where H_n is the set of $\tau \in H$ of order divisible by n. See Exercise 15.4 Now we apply Proposition 15.3 to the field $\mathbf{Q}$. This is Dirichlet's theorem on primes in arithmetic progressions. It ensures that for every $k \geq 0$ there exists a prime number $q \equiv 1 \pmod{n^k}$ that is unramified in K. By Exercise 15.5, the number of elements in $H \cong (\mathbf{Z}/q\mathbf{Z})^*$ of order divisible by n^k is

$$(q - 1)\prod_{p|n}\left(1 - \frac{1}{p^{(k-1)\mathrm{ord}_p(n)+1}}\right).$$

This implies that for every $k \geq 0$, we have

$$\frac{\#H_n}{\#H} \geq \prod_{p|n}\left(1 - \frac{1}{p^k}\right).$$

Now the main result of this chapter follows. By letting k tend to infinity, we see that

$$d_F^-(S) \geq \frac{1}{\#G}.$$

Applying this to all *other* elements of the group G, one finds that the *upper density* $d_F^+(S)$ is *at most* $1/\#G$. Therefore, the lower and the upper densities coincide, and we have

$$d_F(S) = \frac{1}{\#G}.$$

This completes the proof of the theorem.

Theorem 15.6 *(Chebotarëv's density theorem) Let F be a number field and $F \subset K$ be a finite Galois extension. Set $G = \mathrm{Gal}(K/F)$. Then for every conjugacy class C of G, the set S of prime ideals $\mathfrak{p}$ of F whose Frobenius conjugacy class is C has density given by*

$$d_F(S) = \frac{\#C}{\#G}.$$

Proof Let $\sigma \in C$, and let $E \subset K$ be the fixed field of σ. Then K is a cyclic Galois extension of E with Galois group $\langle\sigma\rangle$. By Proposition 15.5, the density theorem holds for the extension $E \subset K$. This means that

$$d_E(T) = \lim_{s\downarrow 1}\frac{\sum_{\mathfrak{q}\in T}\frac{1}{N(\mathfrak{q})^s}}{\log(\frac{1}{s-1})} = \frac{1}{f}.$$

Here F is the order of $\sigma \in G$ and T denotes the set of prime ideals $\mathfrak{q}$ of E whose Frobenius element in $\mathrm{Gal}(K/E)$ is equal to σ. We replace T by its subset of primes of degree 1. This does not affect the density $d_E(T)$.

Let $\mathfrak{p}$ be a prime in S. There are $\#G/f$ primes $\mathfrak{q}$ of E lying over $\mathfrak{p}$. Their Frobenius elements $\varphi_\mathfrak{q}$ in $\mathrm{Gal}(K/F)$ are all conjugate and are therefore contained in C. Therefore, for every $\sigma' \in C$, the number of prime ideals $\mathfrak{q}$ of K that lie over $\mathfrak{p}$ and whose Frobenius element is σ' is *the same*. It follows that this number is equal to $\#G/(f\#C)$. This holds in particular for σ itself. Since for every prime in E over $\mathfrak{p}$ there is precisely *one* prime $\mathfrak{q}$ of K lying over it, we see that there are also $\#G/(f\#C)$

prime ideals of E lying over $\mathfrak{p}$ with Frobenius element equal to σ. From this and the fact that $\mathfrak{p}$ splits completely in E, we deduce that

$$\sum_{\mathfrak{p}\in S} \frac{1}{N(\mathfrak{p})^s} = \frac{f\#C}{\#G} \sum_{\mathfrak{q}\in T} \frac{1}{N(\mathfrak{q})^s}.$$

Letting s tend to 1, we find

$$d_F(S) = \frac{f\#C}{\#G} d_E(T) = \frac{f\#C}{\#G} \cdot \frac{1}{f} = \frac{\#C}{\#G},$$

as required.

Exercises

15.1 Let m be a natural number and let F be a number field containing the mth roots of unity. Let $\mathfrak{p}$ be a prime ideal of F that does not divide m.

(a) Show that the reduction map $\mu_m \longrightarrow (O_F/\mathfrak{p})^*$ is an injective group homomorphism.

(b) Deduce that m divides $N(\mathfrak{p}) - 1$.

15.2 Let $K = \mathbf{Q}(\zeta_3, \sqrt[3]{2})$. The Galois group G of K over $\mathbf{Q}$ is generated by the automorphisms τ and σ of order 2 and 3, respectively. It is isomorphic to the symmetric group S_3. Here τ is given by $\tau(\zeta_3) = \zeta_3^{-1}$ and $\tau(\sqrt[3]{2}) = \sqrt[3]{2}$, while σ is given by $\sigma(\zeta_3) = \zeta_3$ and $\sigma(\sqrt[3]{2}) = \zeta_3\sqrt[3]{2}$. The ring of integers of K is given by $O_K = \mathbf{Z}[\zeta_3, \sqrt[3]{2}]$.

(a) Show that the primes of K that lie over 7 are the ideals $\mathfrak{q} = (\zeta_3 - 2)$ and $\mathfrak{q}' = (\zeta_3^{-1} - 2)$. Show that $G_\mathfrak{q} = G_{\mathfrak{q}'}$ is equal to $\langle\sigma\rangle$ and that the Frobenius element $\varphi_\mathfrak{q}$ is equal to σ^2 while $\varphi_{\mathfrak{q}'} = \sigma$. The Frobenius conjugacy class of the prime 7 is therefore equal to the conjugacy class $\{\sigma, \sigma^{-1}\}$ of G.

(b) Show that the primes of K that lie over 5 are the ideals $\mathfrak{q} = (\sqrt[3]{2} - 3)$, $\mathfrak{q}' = (\zeta_3\sqrt[3]{2} - 3)$, and $\mathfrak{q}'' = (\zeta_3^{-1}\sqrt[3]{2} - 3)$. Show that the groups G_τ, $G_{\tau'}$, and $G_{\tau''}$ are equal to the cyclic groups of order 2 generated by τ, $\tau\sigma$, and $\tau\sigma^{-1}$, respectively. Determine the Frobenius elements of $\mathfrak{q}'$, $\mathfrak{q}'$, and $\mathfrak{q}''$. Show that the Frobenius conjugacy class of the prime 5 is the conjugacy class $\{\tau, \tau\sigma, \tau\sigma^{-1}\}$ of G.

15.3 Let $F = \mathbf{Q}(\sqrt{2}, \sqrt{3})$. The Galois group $G = \mathrm{Gal}(F/\mathbf{Q})$ is generated by σ and τ, where $\sigma(\sqrt{2}) = \sqrt{2}$ and $\sigma(\sqrt{3}) = -\sqrt{3}$, while $\tau(\sqrt{2}) = -\sqrt{2}$ and $\tau(\sqrt{3}) = \sqrt{3}$. The group G is isomorphic to the Klein 4-group. The ring of integers of F is given by $O_F = \mathbf{Z}[\sqrt{2}, \sqrt{3}]$.

(a) Show that the primes of F that lie over 5 are the ideals $\mathfrak{q} = (\sqrt{6} - 1)$ and $\mathfrak{q}' = (\sqrt{6} + 1)$. Show that $G_\mathfrak{q} = G_{\mathfrak{q}'} = \langle \sigma\tau \rangle$. Determine the Frobenius elements of $\mathfrak{q}'$ and $\mathfrak{q}'$.

(b) Show that the primes of F that lie over 7 are the ideals $\mathfrak{r} = (\sqrt{2} - 3)$ and $\mathfrak{r}' = (\sqrt{2} + 3)$. Show that $G_\mathfrak{q} = G_{\mathfrak{q}'} = \langle \sigma \rangle$. Determine the Frobenius elements of $\mathfrak{q}'$ and $\mathfrak{q}'$.

15.4 Let a_n, b_n be two sequences of real numbers. Show that

$$\liminf_{n\to\infty}(a_n + b_n) \geq \liminf_{n\to\infty} a_n + \liminf_{n\to\infty} b_n,$$

$$\limsup_{n\to\infty}(a_n + b_n) \leq \limsup_{n\to\infty} a_n + \limsup_{n\to\infty} b_n.$$

15.5 Let H be a cyclic group of order m and let d be a positive divisor of m. Show that the number of elements in H of order divisible by d is equal to

$$m \prod_{p|d}\left(1 - \frac{1}{p^{a_p - b_p + 1}}\right),$$

where here $a_p = \mathrm{ord}_p(m)$ and $b_p = \mathrm{ord}_p(d)$.

15.6 Let F be a number field.

(a) Show that the series

$$\sum_{\mathfrak{p}}\left(\log(1 - \frac{1}{N(\mathfrak{p})}) + \frac{1}{N(\mathfrak{p})}\right)$$

converges absolutely. Here the summation runs over the prime ideals $\mathfrak{p}$ of F.

(b) Show that the series

$$\sum_{\deg(\mathfrak{p})\geq 2} \frac{1}{N(\mathfrak{p})}$$

converges absolutely.

15.7 Let m be a natural number with $m \not\equiv 2 \pmod 4$, let H be a subgroup of $(\mathbf{Z}/m\mathbf{Z})^*$, and let $H^\vee$ denote its group of characters $\chi : H \longrightarrow \mathbf{C}^*$.

(a) Let $h \in H$. Show that $\sum_{\chi\in H^\vee} \chi(h) = \#H$ when $h = 1$, while it is 0 otherwise. Let $\chi \in H^\vee$. Show that $\sum_{h\in H} \chi(h) = \#H$ when $\chi = 1$, while it is 0 otherwise.

(b) The *conductor* of a character $\chi \in H^\vee$ is the smallest positive divisor f of m for which $\{x \in H : x \equiv 1 \pmod f)\}$ is contained in the kernel of χ. Determine the number of characters of conductor m for $H = (\mathbf{Z}/m\mathbf{Z})^*$.

15.8 (Dirichlet's criterion) Let $a_n \in \mathbf{C}$ be a sequence with the property that $\sum_{n=1}^{N} a_n$ is bounded as $N \to \infty$ and let $b_n \in \mathbf{R}$ be a sequence that tends monotonically to 0 as $n \to \infty$. Show that $\sum_{n=1}^{\infty} a_n b_n$ converges.

15.9 Let $1 \le a \le 9$ be a natural number.

(a) Show that the set S_a of prime numbers whose first (i.e., most significant) decimal digit is equal to a does not have a natural density.

(b) Show that the Dirichlet density of S_a is given by

$$d_{\mathbf{Q}}(S_a) = \frac{\log(a+1) - \log(a)}{\log(10)}.$$

(Hint: Use the prime number theorem)

15.10 Let G be a finite group and let $\sigma \in G$ have order f. Let C denote the conjugacy class of σ. Show that $\#C \cdot f$ divides $\#G$.

15.11 Let $K = \mathbf{Q}(\zeta_3, \sqrt[3]{2})$. Determine the Frobenius conjugacy classes of the Galois group of K over $\mathbf{Q}$ of the primes p satisfying $5 \le p < 50$.

16
Thaine's Theorem

In this chapter we prove an important special case of Thaine's theorem [48].

Let p be an odd prime. Let ζ_p denote a primitive pth root of unity and let $\zeta_p^+ = \zeta_p + \zeta_p^{-1}$. We adopt the convention of chapter 14. This means that we write Cl and Cl^+ for the ideal class groups Cl_p of $\mathbf{Q}(\zeta_p)$ and Cl_p^+ of $\mathbf{Q}(\zeta_p^+)$, respectively. We write for E for the group E_p of p-units in $\mathbf{Q}(\zeta_p)$ and write C for its subgroup C_p of cyclotomic p-units.

Let $G = \mathrm{Gal}(\mathbf{Q}(\zeta_p)/\mathbf{Q})$ and $\iota \in G$ denote the automorphism that induces complex conjugation. We write $F = \mathbf{Q}(\zeta_p^+)$ and let $G^+ = G/\langle \iota \rangle$. We have $G^+ = \mathrm{Gal}(F/\mathbf{Q})$. By E^+ and C^+, we denote the groups of ι-invariant p-units and ι-invariant cyclotomic p-units, respectively. By Exercise 16.1, the natural homomorphism

$$E^+/C^+ \longrightarrow E/C$$

is an isomorphism of G^+-modules.

We have the following "plus" version of Proposition 7.5. As usual, h_p^+ denotes the order of the ideal class group Cl^+ of $\mathbf{Q}(\zeta_p^+)$.

Proposition 16.1 *Let p be an odd prime. Then the group E/C is finite of order h_p^+.*

Proof This result goes back to Kummer. Since we do not need it, we merely sketch the proof. See [50, Ch. 4] for more details. The key point is the fact that the zeta function $\zeta_F(s)$ is a direct product of the L-series $L(s, \chi)$, where χ runs over the characters of G^+. For the nontrivial characters χ of $G = \mathrm{Gal}(F/\mathbf{Q})$, we have

$$L(1, \chi) = -\frac{1}{\tau(\chi^{-1})} \sum_{a=1}^{p-1} \chi^{-1}(a) \log |1 - e^{\frac{2\pi i a}{p}}|,$$

while for $\chi = 1$, the L-series $L(s, \chi)$ is equal to the Riemann zeta function and has a simple pole of order 1 at $s = 1$. Confronting this with the formula for the residue of $\zeta_F(s)$ in $s = 1$ and using the fact that the Gaussian sum $\tau(\chi^{-1})$ has absolute value $\sqrt{p}$, we find

$$h_p^+ \cdot R \;\; = \;\; \prod_{\chi \neq 1} \frac{1}{2} \sum_{a=1}^{p-1} \chi^{-1}(a) \log |1 - e^{\frac{2\pi i a}{p}}|.$$

Here R denotes the *regulator* of F. Like the regulator, the right-hand side of this formula is also equal to a determinant of logarithms of absolute values of units. In this case, cyclotomic units are involved. The quotient of the two determinants is equal to the index $[E^+ : C^+] = [E : C]$.

This proves the proposition.

It is known that the finite groups Cl^+ and E/C are, in general, *not* isomorphic as $\mathbf{Z}[G^+]$-modules or even as groups. Indeed, for $p = 62501$, the 3-parts of Cl^+ and E/C are non-isomorphic groups [50, p. 146]. However, it follows from the results of Mazur and Wiles [30] and of Kolyvagin [22] that for any prime q *not* dividing $p - 1$, the q-parts of Cl^+ and E/C are *Jordan–Hölder isomorphic* $\mathbf{Z}_q[G^+]$-modules [50, Thm. 15.7]. Using the terminology of chapter 13, this is equivalent to saying that for every character χ of G^+, the χ-parts of the q-parts of Cl^+ and E/C have the same order. Thaine's theorem [48] complements this statement. It says that the $\mathbf{Z}[G^+]$-annihilator of the q-part of E/C *also* annihilates the q-part of Cl^+.

Fig. 16.1 Francisco Thaine. (Reproduced with the kind permission of Francisco Thaine).

We prove a somewhat weaker version of Thaine's theorem. The following lemma is the key ingredient.

Lemma 16.2 *We use the notation introduced above. Let p and q be distinct odd primes with q not dividing p − 1. Let C be a class in Cl^+/Cl^{+q}. Then there exists a prime ideal $\mathfrak{l}$ of $F = \mathbf{Q}(\zeta_p^+)$ of degree 1 with the following properties:*

- *$\mathfrak{l}$ lies over a prime number l that is congruent to 1 (mod pq);*
- *the class of $\mathfrak{l}$ in Cl^+/Cl^{+q} is C;*
- *the natural map*

$$E^+/E^{+q} \longrightarrow (O_F/lO_F)^*/(O_F/lO_F)^{*q}$$

is an isomorphism of $\mathbf{F}_q[G^+]$-modules.

Before proving the lemma, we show how it implies Thaine's theorem.

Theorem 16.3 *(F. Thaine, 1988) Let p and q be distinct odd primes with q not dividing p − 1. Let G^+ be the Galois group of $F = \mathbf{Q}(\zeta_p^+)$ over $\mathbf{Q}$. If $\theta \in \mathbf{Z}[G^+]$ annihilates E/CE^q, then it also annihilates Cl^+/Cl^{+q}.*

Proof Suppose that θ annihilates E/CE^q and let $c \in Cl^+$. We have to show that θ annihilates C in the group Cl^+/Cl^{+q}. Let $\mathfrak{l}$ be a prime of $F(\zeta_q)$ in the class C, that has the three properties listed in Lemma 16.2.

Since q does not divide $p − 1$, the ring $\mathbf{F}_q[G^+]$ is semisimple. By Corollary 13.7, the group E^+/E^{+q} is a free $\mathbf{F}_q[G^+]$-module of rank 1. Let $u \in E^+$ be a generator. By Exercise 16.1, the natural map $E^+/E^{+q} \longrightarrow E/E^q$ is an $\mathbf{F}_q[G^+]$-isomorphism. Therefore, u also generates E/E^q as well as its quotient $E/CE^q \cong E^+/C^+E^{+q}$. We have

$$u^\theta = \gamma v^q \quad \text{for some } \gamma \in C^+ \text{ and } v \in E^+.$$

Claim There exists a unit $\varepsilon \in \mathbf{Q}(\zeta_p, \zeta_l)$ for which

$$N(\varepsilon) = 1,$$
$$\varepsilon \equiv \gamma \pmod{(\zeta_l - 1)}.$$

Here the norm is taken from $\mathbf{Q}(\zeta_p, \zeta_l)$ to $\mathbf{Q}(\zeta_p)$.

Indeed, when we have $\gamma = \zeta_p^a - 1$ for some $a \in \mathbf{Z}$, just take $\varepsilon = \zeta_p^a - \zeta_l$. It clearly has the property $\varepsilon \equiv \gamma \pmod{(\zeta_l - 1)}$. Since $l \equiv 1 \pmod{p}$, we have

$$N(\varepsilon) = \frac{\zeta_p^{al} - 1}{\zeta_p^a - 1} = 1,$$

so that ε has the second property as well. Since any $\gamma \in C$ is a product of the p-units $\zeta_p^a - 1$, we can take for ε suitable products of the units $\zeta_p^a - \zeta_l$.

Let τ denote a generator of $\mathrm{Gal}(\mathbf{Q}(\zeta_p, \zeta_l)/\mathbf{Q}(\zeta_p))$. It fixes ζ_p and raises ζ_l to the sth power, where s is some primitive root modulo l. By Hilbert's Theorem 90, there exists $\alpha \in \mathbf{Q}(\zeta_p, \zeta_l)^*$ with the property that $\tau(\alpha)/\alpha = \varepsilon$. See Exercise 16.2. The ideal generated by α is τ-invariant. The extension $\mathbf{Q}(\zeta_p) \subset \mathbf{Q}(\zeta_p, \zeta_l)$ is unramified outside the primes that lie over l. Moreover, the primes that lie over l are totally ramified. Since the primes over l are conjugate under the action of $G = \mathrm{Gal}(\mathbf{Q}(\zeta_p, \zeta_l)/\mathbf{Q}(\zeta_l))$, we have the following equality of $\mathbf{Q}(\zeta_p, \zeta_l)$-ideals:

$$(\zeta_l - 1) = \prod_{\sigma \in G} \sigma(\mathfrak{L}).$$

Here $\mathfrak{L}$ denotes a prime of $\mathbf{Q}(\zeta_p, \zeta_l)$ over l.
　　Exercise 16.4 implies

$$(\alpha) = J \prod_{\sigma \in G} \sigma(\mathfrak{L})^{r_\sigma}. \qquad (*)$$

Here the exponents r_σ are in $\mathbf{Z}$ and J is a $\mathbf{Q}(\zeta_p)$-ideal that is prime to l. The coefficients r_σ are related to the cyclotomic unit γ as follows. For each $\sigma \in G^+$, consider the element $\alpha/(\zeta_l - 1)^{r_\sigma}$. Its denominator does not involve the prime ideal $\sigma(\mathfrak{L})$. Since $\sigma(\mathfrak{l})$ is totally ramified in $F(\zeta_q)$, the automorphism τ acts trivially modulo $\sigma(\mathfrak{L})$. Therefore, we have

$$\frac{\alpha}{(\zeta_l - 1)^{r_\sigma}} \equiv \tau\left(\frac{\alpha}{(\zeta_l - 1)^{r_\sigma}}\right) \equiv \frac{\varepsilon\alpha}{(\zeta_l^s - 1)^{r_\sigma}} \pmod{\sigma(\mathfrak{L})}.$$

Since $\frac{\zeta_l^s - 1}{\zeta_l - 1} \equiv s \pmod{\sigma(\mathfrak{L})}$, we obtain

$$\frac{\alpha}{(\zeta_l - 1)^{r_\sigma}} \equiv \frac{\varepsilon\alpha}{(\zeta_l - 1)^{r_\sigma}} s^{-r_\sigma} \pmod{\sigma(\mathfrak{L})}.$$

Therefore, we have

$$\gamma \equiv \varepsilon \equiv s^{r_\sigma} \pmod{\sigma(\mathfrak{L})}$$

for every $\sigma \in G$. Since both γ and s^{r_σ} are contained in the field F, we find

$$\gamma \equiv s^{r_\sigma} \pmod{\sigma(\mathfrak{l})}$$

for all $\sigma \in G$. Since ι fixes the prime ideal $\mathfrak{l}$, we have the same congruence for $\iota(\gamma)$. It follows that $r_\sigma = r_{\iota\sigma}$ for all $\sigma \in G$. This shows that r_σ depends only on the image of $\sigma \in G^+$.

Since l splits completely in F, the ring O_F/lO_F is the product over $\sigma \in G^+$ of the residue fields $k_{\sigma(\mathfrak{l})}$ of the primes $\sigma(\mathfrak{l})$. Therefore, the $\mathbf{F}_q[G^+]$-module

$$(O_F/lO_F)^*/(O_F/lO_F)^{*q} \cong \prod_{\sigma \in G^+} k_{\sigma(\mathfrak{l})}^*/k_{\sigma(\mathfrak{l})}^{*q}$$

is free of rank 1. Since the residue fields $k_{\sigma(\mathfrak{l})}$ are isomorphic to $\mathbf{F}_l$, the image of the cyclotomic unit γ is equal to $\sum_\sigma r_\sigma \sigma$ applied to a suitable $\mathbf{F}_q[G^+]$-generator $\tilde{u}$. By the third property of the prime ideal $\mathfrak{l}$, the image of the element u *also* generates the rank 1 free $\mathbf{F}_q[G^+]$-module $(O_F/lO_F)^*/(O_F/lO_F)^{*q}$. Therefore, we have $\tilde{u} = u^\eta$ for some unit η of the group ring $\mathbf{F}_q[G^+]$. It follows that

$$u^\theta = \gamma = u^{\eta \sum_\sigma r_\sigma \sigma}$$

in $(O_F/lO_F)^*/(O_F/lO_F)^{*q}$. This implies that $\eta \sum_\sigma r_\sigma \sigma = \theta$ in the group ring $\mathbf{F}_q[G^+]$.

Taking norms from $\mathbf{Q}(\zeta_p, \zeta_l)$ to F of the relation (∗), we find the following equality of F-ideals:

$$(N(\alpha)) = J^{l-1} \prod_{\sigma \in G^+} \sigma(\mathfrak{l})^{r_\sigma}$$

for some F-ideal J. It follows that the element $\sum_{\sigma \in G^+} r_\sigma \sigma$ annihilates the class c of $\mathfrak{l}$ in the group Cl^+/Cl^{+q}. Therefore, $\theta = \eta \sum_\sigma r_\sigma \sigma$ also annihilates the class c in the group Cl^+/Cl^{+q}. This proves the theorem.

The proof of Lemma 16.2 is an application of class field theory and Chebotarëv's density theorem. We recall the following basic fact from class field theory. We do not provide any proof of this result here. Readers interested in the proof are advised to study class field theory [10, 24].

Fact from class field theory Let K be a number field and $\overline{K}$ denote an algebraic closure. The *Hilbert class field* $H(K)$ of K is the maximal extension of K inside $\overline{K}$ that is *abelian* and *unramified at all primes of K*. The Hilbert class field is a *finite* extension of K. It has the property that the map

$$Cl_K \longrightarrow \mathrm{Gal}(H(K)/K),$$

which sends the class of a prime ideal $\mathfrak{p}$ to its Frobenius element $\varphi_\mathfrak{p}$ in the group $\mathrm{Gal}(H(K)/K)$, is a well-defined isomorphism of groups.

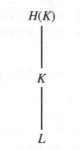

$$H(K)$$

$$K$$

$$L$$

if the field K is itself a Galois extension of a number field L with $\mathrm{Gal}(K/L) = \Delta$, then the class group Cl_K is a $\mathbf{Z}[\Delta]$-module. From the maximality of the Hilbert class field, it follows that $L \subset H(K)$ is a Galois extension so that the group $\mathrm{Gal}(H(K)/K)$ acquires the structure of a Δ through conjugation. In this situation, the homomorphism

$$Cl_K \longrightarrow \mathrm{Gal}(H(K)/K)$$

is a $\mathbf{Z}[\Delta]$-isomorphism.

Examples Let $K = \mathbf{Q}(\zeta_{23})$. Then $H(K) = K(\alpha)$, where α satisfies $\alpha^3 - \alpha + 1 = 0$. The Hilbert class field of $\mathbf{Q}(\zeta_{29})$ is the degree 8 extension of $\mathbf{Q}(\zeta_{29})$ generated by the square roots of zeroes of the polynomial $x^7 + 13x^6 + 31x^5 - 9x^4 - 54x^3 - 3x^2 + 23x - 1$.

We set up the notation that we use in the proof of Lemma 16.2. Suppose p and q are odd primes with q not dividing $p - 1$ and consider the diagram of number fields below. The restriction map

$$\mathrm{Gal}(F(\zeta_q)/\mathbf{Q}(\zeta_q)) \longrightarrow G^+$$

is an isomorphism of groups. Let Δ denote $\mathrm{Gal}(F(\zeta_q)/F)$. Then the Galois group $\mathrm{Gal}(F(\zeta_q)/\mathbf{Q})$ is isomorphic to $G^+ \times \Delta$. Since the group $E^+ \subset F(\zeta_q)^*$ is stable under the action of $G^+ \times \Delta$, the field $F(\zeta_q, \sqrt[q]{E^+})$ is a Galois extension of $\mathbf{Q}$. By Exercise 16.3, the natural map from E^+/E^{+q} to the group $F(\zeta_q)^*$ modulo qth powers is injective. Therefore, Kummer theory implies that the homomorphism

$$\mathrm{Gal}(F(\zeta_q, \sqrt[q]{E^+})/F(\zeta_q)) \longrightarrow \mathrm{Hom}(E^+/E^{+q}, \mu_q)$$

given by $\sigma \mapsto g_\sigma$ where $g_\sigma(\varepsilon) = \sigma(\varepsilon)/\varepsilon$, is an isomorphism. The group $G^+ \times \Delta$ acts on $M = \mathrm{Gal}(F(\zeta_q, \sqrt[q]{E^+})/F(\zeta_q))$ by conjugation. On the other hand, it acts on $\mathrm{Hom}(E^+/E^{+q}, \mu_q)$ via

$$\sigma(f)(a) = \sigma(f(\sigma^{-1}(a))) \text{ for all } \sigma \in G^+ \times \Delta \text{ and } a \in S.$$

These two actions are compatible. In other words, the "Kummer" isomorphism that we describe here is an isomorphism of $\mathbf{Z}[G^+ \times \Delta]$-modules. See also Exercise 9.9. It follows that the group G^+ acts on $\mathrm{Gal}(F(\zeta_q, \sqrt[q]{E^+})/F(\zeta_q))$ through its natural action on E^+/E^{+q}, while Δ acts on $\mathrm{Gal}(F(\zeta_q, \sqrt[q]{E^+})/F(\zeta_q))$ through its action on group μ_q of qth roots of unity. We have $\delta(\zeta_q) = \zeta_q^{\omega(\delta)}$ for $\delta \in \Delta$. Here $\omega : \Delta \longrightarrow (\mathbf{Z}/q\mathbf{Z})^*$ is called the *Teichmüller* character.

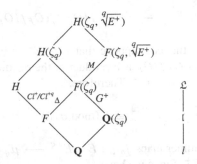

Let H denote the subfield of the Hilbert class field $H(F)$ that corresponds via Galois theory to the quotient group Cl^+/Cl^{+q} of $Cl^+ \cong \mathrm{Gal}(H(F)/F)$. It is the maximal abelian extension of F of exponent q that is contained in $H(F)$. The field $H(\zeta_q)$ is a Galois extension of $\mathbf{Q}$ and $G^+ \times \Delta$ acts by conjugation on $\mathrm{Gal}(H(\zeta_q)/F(\zeta_q)) \cong \mathrm{Gal}(H/F)$. This means that $\Delta = \mathrm{Gal}(F(\zeta_q)/F)$ acts *trivially*. Since Δ acts through the Teichmüller character on $M = \mathrm{Gal}(F(\zeta_q, \sqrt[q]{E^+})/F(\zeta_q))$, it follows that $H(\zeta_q)$ and $F(\zeta_q, \sqrt[q]{E^+})$ are linearly disjoint extensions of $F(\zeta_q))$. Therefore, we have the following isomorphism of $\mathbf{Z}[\Delta \times G^+]$-modules:

$$\mathrm{Gal}(H(\zeta_q, \sqrt[q]{E^+})/F(\zeta_q)) \cong \mathrm{Gal}(H/F) \times \mathrm{Gal}(F(\zeta_q \sqrt[q]{E^+})/F(\zeta_q))$$
$$\cong Cl^+/Cl^{+q} \times M.$$

The quadratic extension $\mathbf{Q}(\zeta_p)$ is linearly disjoint from $H(\zeta_q \sqrt[q]{E^+})$ over F. Therefore, the Galois group of $H(\zeta_p, \zeta_q \sqrt[q]{E^+})$ over $F(\zeta_q)$ is naturally isomorphic to the product of $\mathrm{Gal}(H(\zeta_q \sqrt[q]{E^+})/F(\zeta_q))$ and $\mathrm{Gal}(\mathbf{Q}(\zeta_p)/F) = \langle \iota \rangle$.

Proof of Lemma 16.2 Since q does not divide $p-1$, the ring $\mathbf{F}_q[G^+]$ is semisimple. By Corollary 13.7, the group E^+/E^{+q} is a free $\mathbf{F}_q[G^+]$-module of rank 1. By Exercise 16.5, same is true for the $\mathbf{F}_q[G^+]$-module $M = \mathrm{Hom}(E^+/E^{+q}, \mu_q)$. Let m be a generator of the $\mathbf{F}_q[G^+]$-module M and consider the element

$$(c, m, \ \mathrm{id}\ ) \in \mathrm{Gal}\ (H(\zeta_q, \sqrt[q]{E^+})/F(\zeta_q)) \cong Cl^+/Cl^{+q} \times M \times \langle \iota \rangle.$$

By Chebotarëv's density theorem (Theorem 15.4), the set of primes $\mathfrak{L}$ of $F(\zeta_q)$ whose Frobenius element is equal to (c, m, id) has positive density. Therefore, there exists a prime ideal $\mathfrak{L}$ of $F(\zeta_q)$ of degree 1 with this property. We take for $\mathfrak{l}$ the prime ideal of F over which $\mathfrak{L}$ lies.

We need to show that $\mathfrak{l}$ has the properties required by Lemma 16.2. Since $\mathfrak{L}$ is of degree 1 in $F(\zeta_q) = \mathbf{Q}(\zeta_p, \zeta_q)$, the prime number l over which $\mathfrak{L}$ lies has the property that $l \equiv 1 \pmod{pq}$. The Frobenius element of the prime $\mathfrak{l}$ in $\mathrm{Gal}(H/F)$ is the restriction of the Frobenius element of $\mathfrak{L}$ in $\mathrm{Gal}(H(\zeta_q)/F(\zeta_q))$. Therefore, it is equal to c.

To check the third property, consider the homomorphism

$$E^+/E^{+q} \longrightarrow (O_F/lO_F)^*/(O_F/lO_F)^{*q}.$$

We claim it is *injective*. Indeed, suppose that $\varepsilon \in E^+$ is in the kernel. Since l is totally split in F, the ring O_F/lO_F is a product of the residue fields $k_{\sigma(\mathfrak{l})}$ and ε is a qth power in $k^*_{\sigma(\mathfrak{l})}$ for every $\sigma \in G^+$. Therefore,

$$\varepsilon^{\frac{N(\mathfrak{l})-1}{q}} \equiv 1 \pmod{\sigma(\mathfrak{l})}$$

for all $\sigma \in G^+$. The Kummer maps $f_{\sigma(\mathfrak{l})} : E^+/E^{+q} \longrightarrow \mu_q$ are given by $f_{\sigma(\mathfrak{l})}(\varepsilon) = \sigma(\sqrt[q]{\varepsilon})/\sqrt[q]{\varepsilon}$ for all $\varepsilon \in E^+$. Since we have

$$\sigma(\sqrt[q]{\varepsilon})/\sqrt[q]{\varepsilon} \equiv \varepsilon^{\frac{N(\mathfrak{l})-1}{q}} \pmod{\sigma(\mathfrak{l})} \quad \text{for all } \varepsilon \in E^+,$$

we see that $f_{\sigma(\mathfrak{l})}(\varepsilon) \equiv 1 \pmod{\sigma(\mathfrak{l})}$ and hence $f_{\sigma(\mathfrak{l})}(\varepsilon) = 1$ for all $\varepsilon \in E^+$. Since $f_{\mathfrak{l}}$ is the Frobenius element associated to the prime ideal $\mathfrak{l}$ and $\sigma(f_{\mathfrak{l}}) = f_{\sigma(\mathfrak{l})}$ for every $\sigma \in G^+$, the element $f_{\mathfrak{l}}$ generates $\mathrm{Hom}(E^+/E^{+q}, \mu_q)$ over the group ring $\mathbf{F}_q[G^+]$. It follows that $f(\varepsilon) = 1$ for *every* homomorphism $f \in \mathrm{Hom}(E^+/E^{+q}, \mu_q)$. This means that ε is a qth power and injectivity follows.

Both groups E^+/E^{+q} and $(O_F/lO_F)^*/(O_F/lO_F)^{*q}$ are $\mathbf{F}_q$-vector spaces of dimension $(p-1)/2$. As a consequence, the injective homomorphism $E^+/E^{+q} \longrightarrow (O_F/lO_F)^*/(O_F/lO_F)^{*q}$ is an isomorphism.

This proves the lemma.

One can prove a theorem that is stronger than Theorem 16.2. We quote the following result, a proof of which can be found in [50, Prop. 15.2].

Theorem 16.4 *Let F be a real abelian number field. Let $G = \mathrm{Gal}(F/\mathbf{Q})$, let E denote the group of units of the ring of integers of F, and let C denote the subgroup of cyclotomic integers as defined in [50, p. 15.2]. Let p be a prime not dividing the index $[F : \mathbf{Q}]$. If $\theta \in \mathbf{Z}[G]$ annihilates the p-part of E/C, then 2θ annihilates the p-part of the class group of F.*

Exercises

16.1 Let p be an odd prime. We use the notation introduced in this chapter.

   (a) Show that the natural homomorphism $E^+/C^+ \longrightarrow E/C$ is an isomorphism of $\mathbf{Z}[G^+]$-modules.

(b) For every odd prime $q \neq p$, show that the natural map $E^+/E^{+q} \longrightarrow E/E^q$ is an isomorphism of $\mathbf{Z}[G^+]$-modules. Similarly, the natural map $C/C^q \longrightarrow C^+/C^{+q}$ is a $\mathbf{Z}[G^+]$-module isomorphism.

16.2 Prove Hilbert's Theorem 90: Suppose that $K \subset L$ is a cyclic Galois extension with $\mathrm{Gal}(L/K)$ generated by τ. Then for every $x \in L$ with $N(x) = 1$, there exists $y \in L$ such that $x = \tau(y)/y$.

16.3 Let p, q be odd primes. Show that the natural map

$$\mathbf{Q}(\zeta_p^+)^*/\mathbf{Q}(\zeta_p^+)^{*q} \longrightarrow \mathbf{Q}(\zeta_p)^*/\mathbf{Q}(\zeta_p)^{*q}$$

is injective.

16.4 Let $K \subset L$ be a Galois extension of number fields. Let $G = \mathrm{Gal}(L/K)$ and let I be a G-invariant L-ideal. Show that

$$I = J \prod_{\mathfrak{p}} \mathfrak{p}^{a_{\mathfrak{p}}}$$

for some K-ideal J. Here $a_{\mathfrak{p}} \in \mathbf{Z}$ and the product runs over the prime ideals $\mathfrak{p}$ that are ramified in the extension $K \subset L$.

16.5 Let Γ be finite abelian group and let k be a field. Establish the "Gorenstein property" of the group ring $k[\Gamma]$: show that $\mathrm{Hom}_k(k[\Gamma], k)$ is a free $k[\Gamma]$-module of rank 1. Here the action of Γ on $\mathrm{Hom}_k(k[\Gamma], k)$ is defined by $g(f)(x) = f(g \cdot x)$ for $g \in \Gamma$ and $x \in k[\Gamma]$.

16.6 Let p and q be odd primes and suppose that l is a prime congruent to 1 (mod q) that splits completely in $F = \mathbf{Q}(\zeta_p^+)$.

(a) Show that the prime ideals of F that divide l are conjugate to one another under the action of $G^+ = \mathrm{Gal}(F/\mathbf{Q})$. Show that their residue fields are all equal to the prime field $\mathbf{F}_l$.

(b) Show that the $\mathbf{F}_q[G^+]$-module

$$\prod_{\mathfrak{l}|l} k_{\mathfrak{l}}^*/k_{\mathfrak{l}}^{*q}$$

is free of rank 1. Here $\mathfrak{l}$ runs over the prime divisors of l in F and $k_{\mathfrak{l}}$ denotes the residue field of $\mathfrak{l}$.

Appendix. Euler's Theorem

Here we present Euler's 1738 proof of the fact that the only nonzero rational solutions to the Diophantine equation $X^2 = Y^3 + 1$ are given by $X = \pm 3$ and $Y = 2$. The proof is by a descent argument.

Fig. A.1 Leonhard Euler (1707–1783)

Euler supposes that X, Y are nonzero rational numbers satisfying $X^2 = Y^3 + 1$, lets $Y = Y' - 1$, and supposes that $Y' = c/b$ for coprime integers $b, c \in \mathbf{Z}$. Clearing denominators, this means that

$$bc(c^2 - 3bc + 3b^2) \quad \text{is a square} > 1.$$

The nontrivial solution corresponds to $c = 3$, $b = 1$. We want to show that it is the only one with $\gcd(b, c) = 1$. Since replacing c by $3c'$ leads to the same condition with b,c replaced by c', b, respectively, it suffices to show that the equation has no solution in nonzero coprime integers b,c with $c \not\equiv 0 \pmod 3$). Euler assumes there is a solution and performs a "descent". This means that from the solution X, Y, he produces a new, smaller solution to the equation. Indeed, following his steps, we see that near the end of his proof he writes that $tu(u^2 - 3ut + 3t^2)$ is a square exceeding 1. In a precise sense, this solution is strictly smaller than the original solution. Since this process cannot continue forever, there cannot be any solution at all and the result follows.

THEOREMA 10

Nullus cubus, ne quidem numeris fractis exceptis, unitate auctus quadratum efficere potest praeter unicum casum, quo cubus est 8.

DEMONSTRATIO

Propositio ergo huc redit, ut $\frac{a^3}{b^3} + 1$ numquam esse possit quadratum praeter casum, quo $\frac{a}{b} = 2$. Quocirca demonstrandum erit hanc formulam $a^3 b + b^4$ numquam fieri posse quadratum, nisi sit $a = 2b$.

Haec autem expressio resolvitur in istos tres factores $b(a + b)(aa - ab + bb)$, qui primo quadratum constituere possunt, si esse posset $b(a + b) = aa - ab + bb$, unde prodit $a = 2b$, qui erit casus, quem excepimus. Pono autem, ut ulterius pergam, $a + b = c$ seu $a = c - b$, qua facta substitutione habebitur

$$bc(cc - 3bc + 3bb),$$

quam demonstrandum est quadratum esse non posse, nisi sit $c = 3b$; sunt autem b et c numeri inter se primi. Hic autem duo occurrunt casus considerandi, prout c vel multiplum est ternarii vel secus; illo enim casu factores c et $cc - 3bc + 3bb$ communem divisorem habebunt 3, hoc vero omnes tres inter se erunt primi.

Sit primo c non divisibile per 3; necesse erit, ut singuli illi tres factores sint quadrata, scilicet b et c et $cc - 3bc + 3bb$ seorsim. Fiat ergo $cc - 3bc + 3bb = \left(\frac{m}{n}b - c\right)^2$; erit

$$\frac{b}{c} = \frac{3nn - 2mn}{3nn - mm}, \quad \text{vel} \quad \frac{b}{c} = \frac{2mn - 3nn}{mm - 3nn},$$

cuius fractionis termini erunt primi inter se, nisi m sit multiplum ternarii. Sit ergo m per 3 non divisibile; erit vel $c = 3nn - mm$ vel $c = mm - 3nn$ et vel $b = 3nn - 2mn$ vel $b = 2mn - 3nn$. At cum $3nn - mm$ quadratum esse nequeat, ponatur $c = mm - 3nn$, quad quadratum fiat radicis $m - \frac{p}{q}n$, hincque oritur $\frac{m}{n} = \frac{3qq + pp}{2pq}$ atque

$$\frac{b}{nn} = \frac{2m}{n} - 3 = \frac{3qq - 3pq + pp}{pq}.$$

Quadratum ergo esset haec formula $pq(3qq - 3pq + pp)$, quae omnino similis est propositae $bc(3bb - 3bc + cc)$ et ex multo minoribus numeris constat. At sit m multiplum ternarii, puta $m = 3k$; erit $\frac{b}{c} = \frac{nn - 2kn}{nn - 3kk}$, unde erit vel $c = nn - 3kk$ vel $c = 3kk - nn$; quia autem $3kk - nn$ quadratum esse nequit, ponatur $c = nn - 3kk$ eiusque radix $n - \frac{p}{q}k$, unde fiet $\frac{n}{k} = \frac{3qq + pp}{2pq}$ seu $\frac{k}{n} = \frac{2pq}{3qq + pp}$ atque

$$\frac{b}{nn} = 1 - \frac{2k}{n} = \frac{pp + 3qq - 4pq}{3qq + pp}.$$

Quadratum ergo essere esse deberet $(pp + 3qq)(p - q)(p - 3q)$. Ponatur $p - q = t$ et $p - 3q = u$; erit $q = \frac{t-u}{2}$ et $p = \frac{3t-u}{2}$ illaque formula abit in hanc $tu(3tt - 3tu + uu)$, quae iterum similis est priori $bc(3bb - 3bc + cc)$.

Restat ergo posterior casus, quo est c multiplum ternarii, puta $c = 3d$, atque quadratum esse debet $bd(bb - 3bd + 3dd)$; quae cum iterum similis sit priori, manifestum est utroque casu evenire non posse, ut formula proposita sit quadratum. Quamobrem praeter cubum 8 alius ne in fractis quidem datur, qui cum unitate faciat quadratum. Q. E. D.

Leonhardi Euleri: Theorematum quorundam arithmeticorum demonstrationes, in *COMMENTATIONES ARITHMETICAE*, Volumen I, Ed. Ferdinand Rudio, Lipsiae et Berolini, B.G. Teubner, MCMXV.

References

1. Atiyah, M. and Macdonald I.: *Introduction to Commutative Algebra*, Addison-Wesley, Reading, MA, 1969.
2. Bennett, C., Blass, J., Glass, A., Meronk, D., and Steiner, R.: Linear forms in the logarithms of three positive integers, *J. de Théorie des Nombres de Bordeaux* **9** (1997), 97–136.
3. Bilu, Y.: Catalan's conjecture (after Mihăilescu), *Sém. Bourbaki Exp.* **909**, Nov. (2002).
4. Bilu, Y.: Catalan without logarithmic forms (after Bugeaud, Hanrot, Mihăilescu), *J. de Théorie des Nombres de Bordeaux* **17** (2005), 69–85.
5. Brinkhuis, J.: Gauss sums and their prime factorization, *L'Enseignement Mathématique* **36** (1990), 39–51.
6. Bugeaud, Y. and Hanrot, G.: Un nouveau critère pour l'équation de Catalan, *Mathematika* **47** (2000), 63–73.
7. Cassels, J.W.S.: On the equation $a^x - b^y = 1$, I, *Am. J. Math.* **75** (1953), 159–163.
8. Cassels, J.W.S.: On the equation $a^x - b^y = 1$, II, *Proc. Cambridge Soc.* **56** (1960), 97–113.
9. Cassels, J.W.S.: *Local Fields*, LMS Student Texts **3**, Cambridge University Press, Cambridge, 1986.
10. Cassels, J.W.S. and Fröhlich, A.: *Algebraic Number Theory*, Academic Press, London, 1967.
11. Catalan, E.: Note extraite d'une lettre adressée à l'éditeur, *J. reine angew. Mathematik* **27** (1844), 192.
12. Catalan, E.: Quelques théorèmes empiriques. In *Mélanges Mathématiques*, Mém. Soc. Royale Sciences de Liège, 2° série, **12** (1885), 42–43.
13. Chein, E.Z.: A note on the equation $x^2 = y^q + 1$, *Proc. AMS* **56** (1976), 42–43.
14. Daems, J.: *Het Vermoeden van Catalan*, Nieuw Archief van de Wiskunde **5** (2004), 221–225.
15. Dickson, L.: *History of the Theory of Numbers II*, Chelsea Publications, New York, 1952.
16. Euler, L.: Theorematum quorundam arithmeticorum demonstrationes, pp. 56–58 in *Commentationes Arithmeticae*, Opere Omnia, Series I, Vol. II, Teubner, 1915.
17. Glass, A., Meronk, D., Okada, T., and Steiner, R.: A small contribution to Catalan's equation, *J. Number Theory* **47** (1994), 131–137.
18. Hyyrö, S: Über das Catalan'sche Problem, *Ann. Univ. Turku Ser. A1* **79** (1964), 3–10.
19. Inkeri, K.: On Catalan's conjecture, *J. Number Theory* **34** (1990), 142–152.
20. Jongmans, F.: *Eugène Catalan, Géomètre sans patrie, Républicain sans république*, Société Belge des Professeurs de Mathématique d'Expression Française, Mons 1996.
21. Ko Chao, On the Diophantine equation $x^2 = y^n + 1$, $xy \neq 0$, *Sci. Sinica* **14** (1965), 457–460.

22. Kolyvagin, V.A.: Euler systems, *The Grothendieck Festschrift II*, 435–483. Birkhäuser, Boston, 1990.
23. Kummer, E.E.: Über die Zerlegung der aus Wurzeln der Einheit gebildeten complexen Zahlen in ihre Primfactoren, *J. reine angew. Mathematik* **35** (1847), 327–367. (Collected Papers I, 211–251.)
24. Lang, S.: *Algebraic Number Theory*, Addison-Wesley, Reading, MA, 1970.
25. Lang, S.: *Cyclotomic Fields*, Graduate Texts in Mathematics, **59**, Springer-Verlag, New York, 1978.
26. Langevin, M.: Quelques applications de nouveaux résultats de Van der Poorten, *Sém. Delange-Pisot-Poitou* **4**, Paris.
27. Lebesgue, V.: Sur l'impossibilité en nombres entiers de l'équation $x^m = y^2+1$, *Nouv. Ann. Math.* **9** (1850), 178–181.
28. Lenstra, H.W. and Stevenhagen, P.: Primes of degree one and algebraic cases of Čebotarev's theorem, *Enseign. Math.* **37** (1991), 17–30.
29. Lenstra, H.W. and Stevenhagen, P.: Chebotarëv and his density theorem, *Math. Intelligencer* **18** (1996), 26–37.
30. Mazur, B. and Wiles, A.: Class fields of abelian extensions of **Q**, *Invent. Math.* **76** (1984), 179–330.
31. McCallum, W.: *Consecutive Perfect Powers*, honours project, University New South Wales, Sydney, 1977, unpublished.
32. Mignotte, M. and Roy, Y.: Catalan's equation has no new solutions with either exponent less than 10651, *J. Exp. Math.* **4** (1995), 259–268.
33. Mignotte, M. and Roy, Y.: Minorations pour l'équation de Catalan, *Comptes Rendues Acad. Sci. Paris* **324** (1997), 377–380.
34. Mignotte, M.: Catalan's equation just before 2000, pp. 247–254 in, Jutila and Metsänkylä, eds. *Proceedings Inkeri Symposium, Turku 1999*, de Gruyter, Berlin, 2001.
35. Mihăilescu, P.: A class number free criterion for Catalan's conjecture, *J. Number Theory* **99** (2003), 225–231.
36. Mihăilescu, P.: Primary cyclotomic units and a proof of Catalan's conjecture, *J. reine angew. Mathematik* **572** (2004), 167–195.
37. Mihăilescu, P.: On the class groups of cyclotomic extensions in presence of a solution to Catalan's equation, *J. Number Theory* **118** (2006), 123–144.
38. Ribenboim, P.: *Catalan's Conjecture*, Academic Press, Boston, 1994.
39. Rudin, W.: *Principles of Mathematical Analysis*, McGraw-Hill, New York, 1953.
40. Runge, C.: Ueber ganzzahlige Lösungen von Gleichungen zwischen zwei veränderlichen, *J. reine angew. Mathematik* **100** (1887), 425–435.
41. Schoof, R.: Minus class groups of cyclotomic fields of prime conductor, *Math. Comp.* **67** (1998), 1225–1245.
42. Schoof, R.: Class groups of real cyclotomic fields of prime conductor, *Math. Comp.* **72** (2003), 913–937.
43. Schoof, R.: Review MR2076124 in S. Friedlander, (eds.), Selected mathematical reviews, *Bull. AMS* **43** (2006), 403–414.
44. Schwarz, W.: A note on Catalan's equation, *Acta Arithmetica* **72** (1995), 277–279.
45. Stanley, R.P.: *Enumerative Combinatorics*, Vol. 2. Cambridge University Press, Cambridge, 1999.
46. Steiner, R.: Class number bounds on Catalan's equation, *Math. Comp.* **67** (1998), 1317–1322.
47. Stickelberger, L.: Über eine Verallgemeinerung der Kreistheilung, *Math. Ann.* **37** (1890), 321–367.
48. Thaine, F.: On the ideal class groups of real abelian number fields, *Ann. Math.* **128** (1988), 1–18.
49. Tijdeman, R.: On the equation of Catalan. *Acta Arith.* **29** (1976), 197–209.
50. Washington, L.C.: *Introduction to Cyclotomic Fields*, 2nd edition, Graduate Texts in Mathematics. **83**, Springer-Verlag, New York, 1997.

51. Washington, L.C.: Stickelberger's theorem for cyclotomic fields, in the spirit of Kummer and Thaine, pp. 990–993 in J.-M. de Koninck and C. Levesque, *Number Theory*, Proc. Int. Conf. Laval 1987, de Gruyter, Berlin, 1989.

52. Weil, A.: *Number Theory, An Approach Through History; from Hammurapi to Legendre*, Birkhäuser, Boston, 1984.

Index